THIRD EDITION

- Bruce Leadbeatter
- Michael Leadbeatter

T0348159

WOODWORKING
Part Two

NELSON
CENGAGE Learning

Australia • Brazil • Japan • Korea • Mexico • Singapore • Spain • United Kingdom • United States

NELSON
CENGAGE Learning

Woodworking Part Two
3rd Edition
Bruce Leadbeatter
Michael Leadbeatter

Publishing manager: Natalie Muir
Production editor: Megan Lowe
Editor: Catherine Page
Cover design: Antart
Design: Antart
Illustrator: Lorenzo Lucia
Reprint: Jess Lovell
Typeset in Rotis Sans Serif by Post Pre-press Group

Any URLs contained in this publication were checked for
currency during the production process. Note, however,
that the publisher cannot vouch for the ongoing currency of
URLs.

First published 2001 by McGraw Hill Australia.
Reprinted 2005, 2007, 2008 by McGraw Hill Australia.
This edition published in 2010 by Cengage Learning
Australia.

Acknowledgements
Additional owners of copyright material are named on the
acknowledgements page.

Text © 2001 Bruce and Michael Leadbeatter
Illustrations and design © 2010 Cengage Learning Australia Pty Limited

For product information and technology assistance,
in Australia call **1300 790 853**;
in New Zealand call **0800 449 725**

For permission to use material from this text or product, please email
aust.permissions@cengage.com

National Library of Australia Cataloguing-in-Publication Data
Leadbeatter, Michael.
Woodworking. Part two / Michael Leadbeatter, Bruce Leadbeatter

9780170198127
Includes bibliographical references and index.
For secondary school & TAFE students.

Woodwork--Textbooks.
Woodwork--Problems, exercises, etc.
Leadbeatter, Bruce R.

684.08076

Cengage Learning Australia
Level 7, 80 Dorcas Street
South Melbourne, Victoria Australia 3205

Cengage Learning New Zealand
Unit 4B Rosedale Office Park
331 Rosedale Road, Albany, North Shore 0632, NZ

For learning solutions, visit **cengage.com.au**

Printed in Australia by Ligare Pty Limited.
4 5 6 7 8 9 10 19 18 17 16 15

CONTENTS

PREFACE

This book is designed to be used in conjunction with *Woodworking, Part One* by the novice woodworker as well as the experienced craftsperson.

During the last decade there have been a number of significant advances in engineered materials, building techniques and the design of portable power tools for general use. Once again, we visited factories and workshops and talked to people on the job, so that we could see for ourselves how these developments have affected the woodworking industry. The response was rather overwhelming and a great deal of time was required to sift and condense the information to usable proportions.

There is no doubt that the growing worldwide interest in woodcraft has been spurred on by the design and marketing of portable power tools. We approached the second chapter with this in mind and expended a great deal of time and effort in trying out and evaluating the potential of these tools for the craftsperson.

We took time off from teaching and lecturing to build a three-level house at Lennox Head in NSW. This was an enjoyable and rewarding exercise which gave us the opportunity to try out most of the new materials and building techniques developed for the industry. We designed and built the roofing trusses, the Californian redwood window sashes and frames and the three stairways. We also tied down the framework of the building to cyclone specifications.

There has been a growing worldwide interest in the craft of woodturning. The craft in many respects has become more creative and is recognised as an art form. This in turn has led to the development and manufacture of new lathes, chucks, tools and accessories to satisfy both the basic and artistic needs of woodturners. We therefore felt it was necessary to add a new chapter, called Advances in Woodturning, to supplement the woodturning chapter in *Part One*. The chapter covers some of the developments that open new horizons for the woodturner.

Although *Woodworking, Part One* and *Woodworking, Part Two* have been designed predominantly as texts for schools and colleges, they were written for all who have a desire to work with or a love of wood.

Two or three generations ago there was talk of the dying interest in woodworking; plastics were the material of the future. Today wood is so commonplace we invariably take it for granted. Our homes and workplaces are at least partially constructed from timber; we eat, sleep and work with wooden furniture; our children play with wooden playthings; and even as adults wood supplies our recreational needs in the form of cabinetmaking and woodturning.

Working with wood reveals a special beauty; it is both warm and pleasing to the touch and its colour and texture are a delight to the eye. The nature of wood is such that it imparts a uniqueness to every single workpiece— something that cannot be said of plastic or metal. As with other woodworkers around the world, we are concerned that many of our valuable and exotic timbers are a diminishing resource. Responsible action must be taken to preserve, protect and replace the world's rainforests.

B.L. M.L.

ACKNOWLEDGMENTS

The authors wish to express their sincere appreciation for the assistance given by the following agencies, organisations and companies in the preparation of this book:

Australian Design Centre

Binks-Bullows (Aust) Pty Limited

Black & Decker (Australasia) Pty Ltd

CSIRO

Carba-Tec

Chiswell Furniture Pty Ltd

Colliers International Limited

Connell Wagner (Laminated Beam)

Forestry Commission of NSW

Hall's Machinery Pty Ltd

Hitachi Sales Australia Pty Ltd

James Hardie & Coy Pty Limited

Makita Australia Pty Ltd

McCulloch of Australia Limited

National Association of Forest Industries

National Trust of Australia

The Plywoods Association of Australia

Presbyterian Ladies College, Sydney

The Regent Sydney

Timber Development Association of Australia (NSW) Ltd

Timber Engineering Co. Pty Ltd

Wattyl Australia Pty Ltd

Woodfast Machinery Co

The Authors also wish to acknowledge: Mr Bob Buckley, School of Building, Sydney Technical College; Roger Gifkin; Greg Koutoulas; Frankie La Spada; and special thanks to many teachers who tendered constructive advice.

For photographs and illustrative material, we thank: Andrew Kay Photography P/L, Figs. 1.2, 1.6, 1.7(a), 1.8(a), 1.9, 1.10, 1.11, 1.16, 1.20, 1.22, 1.23, 1.24, 1.25, 1.27, 1.33, 1.35, 1.36, 1.39, 1.40, 1.41, 1.46, 1.47, 1.48, 1.49, 1.50, 1.51, 1.52, 1.53, 1.54, 1.55, 1.56, 2.3, 2.7, 2.8, 2.10, 2.11, 2.19, 2.20, 2.21, 2.22, 2.25, 2.26, 2.27, 2.28, 2.35, 2.36, 2.37, 2.38, 2.42, 2.46, 3.92(c), 3.92(g), 3.92(h), 3.92(j), 3.92(k), 5.3, 5.4, 5.5, 5.6, 5.7, 5.10, 5.13, 5.15, 5.16, 5.17, 5.18, 5.19, 5.20, 5.21, 5.22, 5.26, 5.27, 5.28, 6.1, 6.14, 6.15, 6.16, 6.18, 6.19, 6.20, 6.27, 9.28, 9.32; Carba-Tec, Fig. 2.34; Covermore Designs, Figs. 7.62, 9.25, 9.38; Craftmaster Products, Fig. 1.43; Department of Arts, Heritage and the Environment, Canberra, Fig. 8.7; *Department of Employment Manual*, Australian Government Publishing Service, 1988, Fig. 2.1; Foresty Branch, Department of Primary Industries and Energy, Fig. 8.1; Mr Daniel Low, Forestry Commission of NSW, Fig. 8.6; Hall's Machinery Pty Ltd, Figs. 7.42, 7.46, 7.47, 7.49; Hitachi, Figs 2.16, 2.17, 2.23, 2.29; VICMARC Machinery Pty. Ltd. Australia, Figs. 5.8, 5.9(a), 5.9(b); *Houses* (2000 : 18) Fig. 7.65; *Houses* (2000 : 21) Fig. 7.63; *Houses* (2000 : 22) Fig. 7.69; Makita, Figs. 2.4, 2.5, 2.6, 2.9, 2.13, 2.14, 2.15, 2.24, 2.30, 2.31(a), 2.31(b), 2.31(c), 2.32, 2.33, 2.41, 2.43, 7.15, 7.41; McCulloch of Australia Limited; Ms Sian Hewitt National Association of Forest Industries Ltd, Figs. 8.4, 8.5; National Trust of Australia (NSW), Figs. 7.70 (J Whitelock, 1966), 7.71 (G Karskens, 1981), 7.73, 7.75 (L G Clark, 1965), 7.76 (Jenny Rix, 2000), 7.77(a) (J Whitelock, 1971), 7.77(b) (L G Clark, 1965), 7.78 (L G Clark, 1966), 7.79, 7.80, 7.81, 7.82, 7.83 (J Whitelock, 1967); Parker Furniture, Figs. 7.64, 7.68; The Regent Hotel, Sydney, Fig. 7.101; Scheppach Maschinenfabrik, Fig. 1.1; Victorian Government, Fig. 8.3.

A very special thanks must go to our wives, Pat and Kerrie, and our families for their patience and understanding during the long hours they were deprived of our company during the preparation of this book.

B.L.

M.L.

GENERAL SAFETY

Fixed or stationary woodworking machines are safe when proper precautions are observed. Specific precautions are included in the sections dealing with individual machines. The following general safety rules apply to all machines throughout the chapter.

1. **Permission.** *Obtain permission from the teacher or instructor before operating any of the machines.*

2. **Specific safety rules.** *Read the instruction manual for the machine you wish to use and make yourself familiar with any special precautions that may be necessary.*

3. **Eye protection.** *Always wear safety glasses or goggles when operating any power machine.*

4. **Hearing protection.** *Always wear earmuffs or earplugs to avoid hearing damage when working on noisy machines.*

5. **Jewellery.** *Metal rings on your fingers can be dangerous if splinters are present on the timber. Remove rings before using any of the machines.*

6. **Clothing.** *Do not wear loose clothing; remove your tie, roll your sleeves up and tie long hair back.*

7. **Dust.** *Protect yourself from the long-term health problems that can result from dust inhalation. Wear a dust mask if an efficient dust extraction system is not available.*

8. **Lighting.** *Make sure that shadowproof lighting is available to provide an unobstructed view.*

9. **Cutters and accessories.** *Never use blunt cutters as they are dangerous. Always check accessories for secure fit before operating a machine.*

10. **Safety guards.** *Make sure that safety guards and hold-down devices are fixed securely before starting a machine.*

11. **Waste material.** *Always remove offcuts, sawdust or shavings from the working surface of machines before turning on the power.*

12. **Starting.** *Always allow a machine to attain maximum speed before starting any process.*

13. **Hand protection.** *Never reach across the path of any form of cutter on a power machine while it is in operation.*

14. **Adjustments.** *Never make adjustments while a machine is operating. Make sure adjustments are secure and check for accuracy on a waste piece if necessary.*

15. **Stopping.** *Never leave a machine running unattended. Turn the power off, and make a point of remaining with the machine until it has come to a complete stop.*

Chapter 1
Woodworking Machines and Processes

BANDSAWS

The bandsaw is one of the most versatile machines in the workshop. It can rip, cut curves, resaw and cut joints. It is reasonably easy to maintain and does not require much floor space.

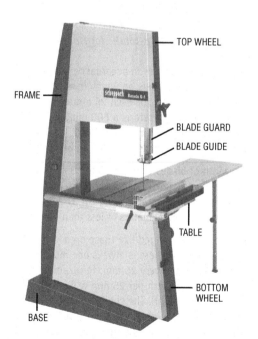

FRAME

TOP WHEEL

BLADE GUARD

BLADE GUIDE

TABLE

BOTTOM WHEEL

BASE

Figure 1.1 A two-wheeled bandsaw

The bandsaw cannot make as smooth a cut as a table saw as the table saw has a stiffer thicker blade that stays straighter in the cut.

However, a bandsaw is best for cutting curves and resawing wide stock with minimal waste because the depth of cut is greater and the blade is narrower.

PARTS AND USES

The main parts of a typical wood-cutting bandsaw are shown in Figure 1.1.

Base—*The metal support for the machine.*

Blade guard—*A sheet metal piece attached near the upper blade guide. It protects the operator by covering the blade from the upper guide to the upper wheel.*

Blade guides—*One guide is situated above, the other below the table (Fig. 1.2). Prevent the blade from twisting, therefore assuring a straight cut.*

Frame—*The metal structure that supports the two wheels and the internal mechanism. The size of the bandsaw is indicated by the width of cut that can be made. For use in the home workshop, the 350 to 750 mm bandsaw is the most convenient size. The smaller 300 to 350 mm bandsaws tend to work harden the blades, causing them to break in a very short time.*

Table—*The platform that supports the work being cut. Most machines allow the table to tilt at 45° to the right and 10° to the left.*

Tension adjustment handle—*Generally located on the upper wheel. The pressure on the blade is controlled by operating the handle.*

Figure 1.2 Plastic or brass guides prevent blade from twisting; ball-bearing takes thrust of blade (guard removed for clarity)

Wheels—*There are usually two wheels (some large throat bandsaws have three wheels) which are covered by rubber tyres. The bandsaw blade runs on these tyres, which protect the teeth and serve as a cushion, preventing the blade from slipping.*

BANDSAW BLADES

There are many types of metal-cutting bandsaw blades available depending on the size and the type of metal being cut; the choices for woodcutting are not as numerous. The type of bandsaw blade used in cutting timber is generally one of three types: the standard tooth, or regular tooth; the skip-tooth, or buttress tooth; and the hook-tooth.

All woodcutting bandsaw blades have each tooth set alternately; a raker (or unset) tooth serves little purpose in cutting wood. For cutting thin timber a regular blade (Fig. 1.3(a)) is sufficient. Every second tooth is eliminated

on the skip-tooth blade (Fig. 1.3(b)) allowing it to clear itself better and cut much faster. The hook-tooth blade (Fig. 1.3(c)) has an increased chip clearance allowing the operator to put a rake angle on the tooth of usually 10°, making feeding easier and cutting faster.

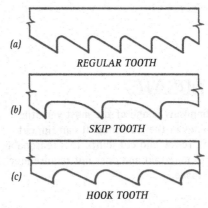

(a)

REGULAR TOOTH

(b)

SKIP TOOTH

(c)

HOOK TOOTH

Figure 1.3 Types of bandsaw blades: (a) Regular, (b) Skip-tooth; (c) Hook-tooth

Blade thickness varies in accordance with the diameter of the wheels on which it runs; generally 0.025 mm for each 25 mm of the wheel diameter. Therefore, a 508 mm diameter wheel needs a 0.508 mm thick blade, however, both thinner and thicker blades can be used.

The width of blades vary from 3 to 19 mm for general workshop use. Width is determined by the diameter of the curve being cut. Use a jigsaw or drill on diameters of less than 13 mm.

The teeth are arranged like those on a handsaw, in that there is always one more point than teeth every 25 mm; for example, a blade with five teeth per 25 mm will have 6 points (Fig. 1.4(b)). The choice of blade depends on the thickness of timber being cut: the thicker the timber, the fewer teeth per 25 mm. For resawing, use two or three teeth per 25 mm, while ten or more teeth should be used for cutting thin timber. A good general purpose blade has around five teeth per 25 mm. The teeth are sharpened like those on a ripsaw blade.

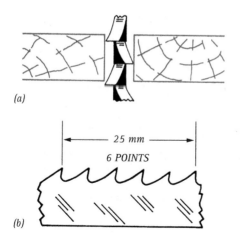

(a)

25 mm

6 POINTS

(b)

Figure 1.4 (a) Bandsaw blades are sharpened with rip teeth (blades with induction-hardened teeth retain their cutting edges much longer); (b) The size of tooth is designated by number of points per 25 mm

Adjusting the blade

Before making any adjustments, make sure the power to the machine is turned off. Loosen the tension adjustment handle so that you can easily move the blade around the wheels and back off the blade guides.

Note: When installing a blade, make sure that the teeth are going in the right direction.

To adjust the blade correctly, you should do things in the right order.

1. Tensioning. *Wind the tensioning handwheel until the correct tension is achieved. Most bandsaws have a tensioning gauge which tells you the correct tension for the width of blade used. If the saw does not have a gauge, you have to develop your own technique—some people pluck the handsaw blade like a guitar and learn to recognise the sound the blade makes at the right tension. Too much tension and the blade will break; too little and the blade will wander in the cut.*

2. Tracking. *Apply enough tension to hold the blade on the wheels, then rotate the wheel by hand and adjust the tracking knob until the blade rides on the centre of the wheels. Now finish tensioning the blade and test track by hand again. At this stage you can close the guard doors and turn the bandsaw on to test the tracking at higher speeds. Turn it on and off quickly at first, then let it run. Never track the blade or open the guard door while the saw is running at high speed.*

3. Squaring the table to the blade. *Use a try square to check that the table is square to the blade. If not, adjust it by loosening the locking lever normally positioned under the table.*

4. Adjusting the blade guides. *Bandsaws generally have two sets of blade guides; one below the table and one above. The top sets up and down for different thicknesses of cut. Each guide consists of a thrust bearing and two side guides. The thrust bearings should be set firstly, then the side blocks. If the guide has ball thrust bearings, they can be brought into contact with the blade, but hardened steel blocks or pivoting blocks need to be set so they are not quite touching the blade (Fig. 1.5). Rotate the saw by hand to test the setting because some bandsaw blades that are not welded exactly straight tend to pulse in use. The blade should just touch in motion as any more than this will wear out the bearings. Now start adjusting the left guide blocks, both top and bottom. These should be adjusted to about a paper width from the blade. As well as adjusting the guides sideways, check that the blocks are in the proper relation to the front of the blade; they should be just behind the gullet of the teeth.*

It is important to check the table for square again to ensure that you are not bending the blade slightly (Fig. 1.6). Now adjust the right-hand blocks to a paper width from the blade. Test the adjustments by moving the blade by hand before switching on the machine.

BLADE
SUPPORT

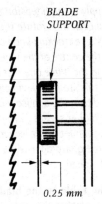

0.25 mm

Figure 1.5 Track blade so that it just clears top and bottom thrust race

Figure 1.6 Check squareness of bandsaw blade to table

Folding the blade

Bandsaw blades should be brushed with a fine wire brush to remove any gummy build-up behind the teeth and folded into three coils for easy storage. They should then be sprayed with an oily rust inhibitor and stored in a plastic bag ready for future use.

Fold the blade by holding it with thumbs pointing up and toe supporting the bottom

(Fig. 1.7(a)), then twist the blade by pointing both thumbs in and past each other (Fig. 1.7(b)). Turn the top loop under and drop it to form three coils.

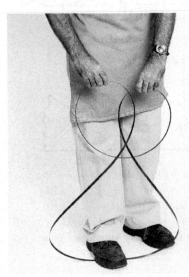

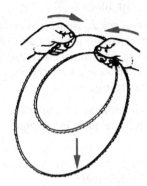

Figure 1.7 Folding bandsaw blade for storage: (a) Grasp blade as shown with teeth facing away from the body; (b) Turn left hand then right hand inwards and drop the blade

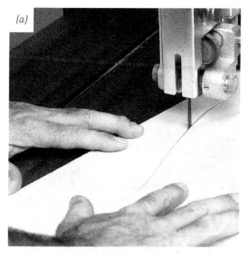

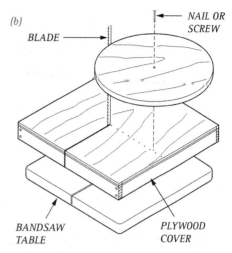

Figure 1.8 (a) Use smooth, uniform action to cut large curves, or (b) improvise pivot jig as shown for cutting circles

CURVE CUTTING

With the use of a jig, the handsaw can cut reasonably accurate curves and circles (Fig. 1.8).

Adjust the top guide to within 12 mm of the job to be cut. While cutting, keep your fingers to the side of the blade or behind it, never in front.

Never use a blade too wide for the radius of the curve being cut as this can put stress on the back of the blade and cause it to break. To obtain a smooth cut you need to feed the timber in evenly. If it is a tight curve (Fig. 1.9), you need to relieve the cut by making straight cuts in the waste to remove most of it.

Note: Never force the blade into the curve if the blade does not want to cut the curve. Cut into the waste and come back to cut to the line after the waste is removed. Never apply force to the blade if you are backing out of a cut (this will force the blade off the wheels).

STRAIGHT CUTTING

Many woodworkers have had bad experiences in using the bandsaw to cut a straight line. The bandsaw, probably more than any other machine, needs a delicate or learned touch.

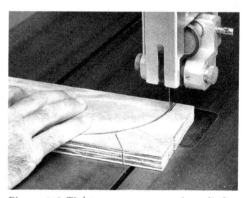

Figure 1.9 Tight curves may require relief cuts to prevent blade from binding

In making a straight cut, it is important to ensure that the machine is adjusted correctly and the guides are set carefully so the blade can run freely yet be supported in its travel. Bandsaw teeth form an area around the blade which must not touch the blade, otherwise any side pressure will twist the blade causing it to cut unevenly. This cutting area must be determined and the fence aligned with it.

Note: The cutting path is rarely parallel to the sides of the table, therefore an adjustable fence is necessary which should be regularly checked with the blade (Fig. 1.11).

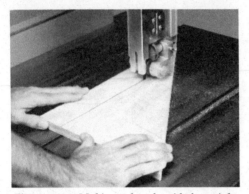

Figure 1.10 Making a hand-guided straight cut, make allowance for run-off of blade

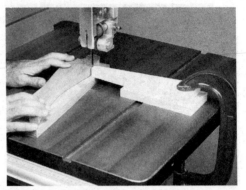

Figure 1.11 Sawing duplicate shapes by using wooden pattern as guide

SAFETY RULES FOR THE BANDSAW

1. **General safety.** *Observe the general safety rules given at the beginning of this chapter.*

2. **Blade size.** *Use the correct blade width for the job in hand.*

3. **Blade condition.** *Do not use a blunt blade. Listen to the blade when it is running; a clicking noise indicates cracking and impending breakage. If the blade breaks, do not open the covers or attempt to remove the blade until the wheels have stopped running.*

4. **Guides.** *Check upper and lower guides for correct alignment and adjust upper guide down to within 10 mm of the stock to be cut.*

5. **Blade lead and tension.** *Check blade tension for compliance with manufacturers' specifications (a middle-C pitch is a rule of thumb). Adjust the blade to track on the top of the crown of the wheels so that it cuts parallel to the table. When the blade is running freely it should exert little or no pressure on the ball-bearing guides.*

6. **Hand protection.** *Never place hands directly in front of the blade. Keep fingers at least 50 mm away from the blade at all times. Use push sticks to support the job if the stock is very small.*

7. **Sawing.** *Freehand sawing should be attempted only when the work to be sawed rests flat on the table. Do not attempt to cut across round stock as it will tend to grab and break the blade. Never attempt to withdraw the stock while the blade is still in the cut. Stop the machine and withdraw the job when the blade has stopped moving.*

8. **Curve cutting.** *Make sure the radius of the cut is not too small for the width of the blade. It may be necessary to make relief cuts to achieve small radii.*

CIRCULAR SAWS

The circular saw was one of the first power machines used in woodworking. Most modern table saws have a tilting arbor (to 45°), a raising and lowering mechanism and provision for dust extraction. Travelling and extension tables are also available where easier handling of large sheets of plywood or particle board is required. The 300 mm blade runs at approximately 3000 rpm (revolutions per minute). The cutting speed, measured in surface metres per minute, is approximately 3000 mpm. The table saw in Figure 1.12 is typical of the machines used in school or college workshops. It has a single tilting arbor and blade driven by an electric motor via twin V belts.

The specified size of a circular saw is measured by the largest diameter saw blade it will accommodate. The saw shown in Figure 1.12 is classed as a 300 mm bench saw, but it can take smaller diameter blades where required.

PARTS AND USES

The main parts of the circular saw are shown in Figure 1.12.

Table—*The top of the saw bench which supports the rip fence, cut-off guide and safety guard.*

Rip fence—*Used as a guide for ripping or cutting parallel to the blade.*

Cut-off guide—*Also called a mitre gauge and used as a guide for crosscutting.*

Arbor—*Holds the blade or dado head in its shaft.*

Saw-tilt handwheel—*A handwheel for tilting the blade to the correct angle for cutting bevels, mitres and right-angle cuts.*

Blade raising wheel—*Regulates the cutting height of the saw blade.*

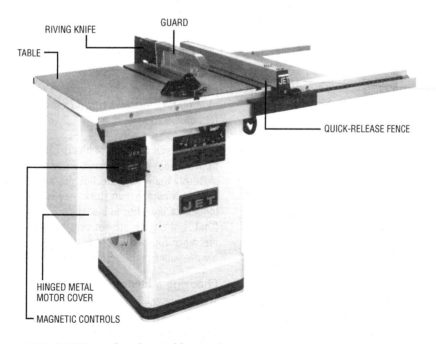

Figure 1.12 A 300 mm bench or table circular saw

Riving knife—Sometimes called a splitter as it is attached to the arbor behind the blade. Its purpose is to prevent the timber closing in on the blade and being thrown back towards the operator.

Safety guard—Designed to protect the operator from the saw and to aid in dust extraction.

Electric motor—Drives the circular saw, usually by means of twin V belts.

CIRCULAR SAW BLADES

There are two types of circular saw blades available: those made of carbon steel alloyed with nickel and chrome and those made of spring steel with tungsten carbide tips brazed on to form the teeth. Alloy steel blades must have their teeth spring set to provide a kerf wider than the thickness of the blade. Alternatively, the blade may be hollow ground so that it does not bind in the cut. There are three basic types of alloy steel blades:

1. *The ripblade is used for cutting thick timber along the grain (Fig. 1.13(a)). The teeth should have a positive rake with a coarse set and be filed or sharpened so that they have a chisel-like action.*

2. *The crosscut blade is used for cutting timber across the grain (Fig. 1.13(b)). The teeth should have a negative rake and be sharpened on alternate sides to produce a knife-like action that cuts the fibres cleanly.*

3. *The combination blade is a general purpose blade designed to cut both along and across the grain and to provide a smoother finish (Fig. 1.13(c)). The hollow-ground blade tends to be smoother than the spring-set blade, mainly because it requires no set and the edges of the deep gullets have a planing action.*

Tungsten carbide blades are available in a wide range of saw-tooth patterns but once again they can all be placed in the rip, crosscut or combination categories (Fig. 1.14).

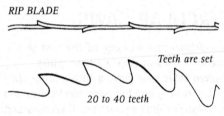

RIP BLADE

Teeth are set

20 to 40 teeth

File teeth faces perpendicular to blade

(a)

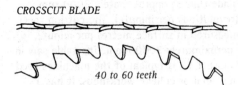

CROSSCUT BLADE

40 to 60 teeth

File teeth faces with alternating bevels

(b)

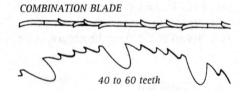

COMBINATION BLADE

40 to 60 teeth

Raker tooth,with deep gullet, clears chips

(c)

Figure 1.13 Three basic types of teeth used on circular saws: (a) rip; (b) crosscut; (c) combination

For any given category of blade, carbide-tipped blades almost always cut more smoothly than sharp steel blades and, because carbide is harder than steel, they stay sharp longer. Carbide-tipped blades do not have set (that is, the teeth are not bent to give clearance for the blade) because the tips are always wider than the thickness of the blade.

Choosing a blade

The carbon steel blades are relatively inexpensive and can be readily reset and sharpened in the workshop, whereas the carbide-tipped blades cost more and have to

CARBIDE COMBINATION

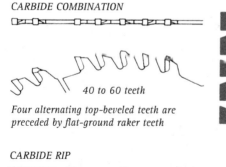

40 to 60 teeth

Four alternating top-beveled teeth are
preceded by flat-ground raker teeth

CARBIDE RIP

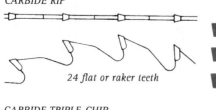

24 flat or raker teeth

CARBIDE TRIPLE-CHIP

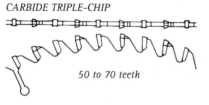

50 to 70 teeth

Triple-chip tooth alternates with raker tooth

*Figure 1.14 Tungsten carbide-tipped
circular saw blades*

be sent out for sharpening. Carbide-tipped
blades, however, produce a smoother cut and
are essential for cutting plywood and particle
board as they hold their cutting edge for much
longer. They are more expensive initially but
more economical in the long run.

A combination carbide-tipped blade with 40 to
60 teeth will give good all-round service for
cutting solid timber, particle board and plywood.
The teeth should be ground with a series of four
alternately bevelled teeth, followed by a flat
raker tooth. For ripping only, a carbide-tipped
ripsaw blade with 24 teeth, positive hook and
flat ground will give excellent service (Fig. 1.15).

Care of blades

Saws are most frequently damaged by
overheating. This can be caused by trying to
force the wood too fast through the saw or by
using a dull blade. Overheating usually warps
the blade and it may be necessary to have it
sent out to be hammered or retensioned.

Resinous timbers have a tendency to gum up
the saw teeth and cause it to burn—so frequent
cleaning (using suitable solvents such as mineral
turpentine or paint stripper) may be necessary.
When carbide-tipped teeth become dull they
require regrinding and of course this shortens
the life of the blade. (Always treat tipped blades
with great care as tungsten carbide is a very
brittle metal subject to chipping.)

The saw blade should run perfectly true on the
arbor and be parallel to the mitre-guide slots
on the table to allow the rip fence to be
correctly aligned. The rip fence may be set
parallel to the blade or a fraction wider on the
outfeed side so that the trailing teeth do not
tear the fibres.

OPERATING ADJUSTMENTS

Manufacturers generally supply service or
operating manuals with their machines. Study
these carefully for information on service and
use.

All cast-iron surfaces and exposed moving
parts should be lightly coated with floor wax
which prevents rust and acts as a lubricant.
Grease or oil stains woodwork and when
combined with wood dust, sets into a
semisolid mass and loses its lubricating ability.

Bearings for woodworking machinery are
usually sealed in the factory to prevent the
entry of dust, and will need to be replaced
from time to time.

Removing and installing a saw blade

1. *Remove the throat plate that surrounds the
   blade.*

2. *Raise the arbor as high as possible.*

3. *Undo the nut and flange by wedging a piece
   of wood between the teeth of the blade and
   the table to lock the blade. Use a well-fitting*

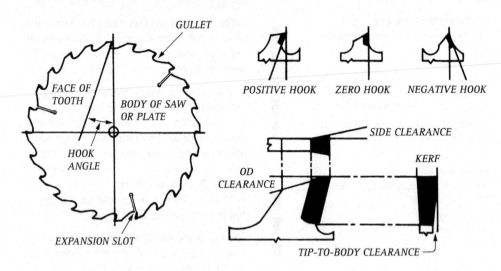

Figure 1.15 Typical chisel or ripsaw tooth

ring spanner and shock load it if necessary with a soft hammer or mallet.

4. Replace the blade by reversing the above three steps, without overtightening the nut.

Adjusting the rip fence

The rip fence may get out of parallel with the blade and may require adjustment from time to time. It usually has two machine screws fixing it to the sliding mechanism. It is simply a matter of loosening these screws and adjusting the fence parallel to the groove on the saw table (Fig. 1.16).

Adjusting the mitre gauge or cut-off guide

Set the cut-off guide at 90° to the blade by using a try square as shown in Figure 1.17.

Adjusting the depth of saw blade

Most table saws have a raising and lowering mechanism as shown in Figure 1.18.

RIPPING

Ripping is the process of sawing timber lengthwise or along the grain. Make sure that the timber to be cut has a straight edge and is not twisted or in wind as kickback could occur.

Never stand in line with the saw blade.

1. Set the ripsaw or combination blade so that it has maximum depth of cut.

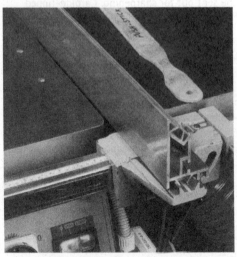

Figure 1.16 Setting fence parallel to blade by lining it up with table groove

Figure 1.17 Use a try square to set cut-off guide at 90° to blade

MAGNETIC CONTROLS

Figure 1.18 Adjusting wheel on left sets height of blade; wheel at right adjusts tilt

2. Set the rip fence to the desired width (Fig. 1.19).

3. Make a test cut on a piece of scrap timber.

4. A riving knife or splitter is a safety must for all ripping. Adjust the saw guard to a safe height and proceed to feed the timber smoothly past the saw (Fig. 1.20).

Figure 1.19 Set width of cut by measuring from fence to edge of blade

5. Long boards should be supported by a table extension or by someone tailing-out. Cutting the board halfway from each end is another alternative.

Figure 1.20 Use feather board and push stick to assist in ripping

Ripping a taper

Make an adjustable jig as shown in Figure 1.21 or a set jig for ripping as shown in Figure 1.22.

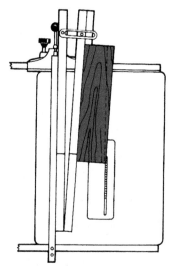

Figure 1.21 Adjustable jig for taper cutting is easily constructed

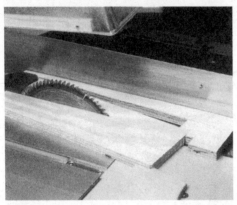

Figure 1.22 Set or fixed taper jig runs against rip fence (guard removed for clarity)

CROSSCUTTING

Crosscutting is the process of sawing timber across the grain. Some of the larger industrial saw benches have a travelling or moving table to facilitate the cutting of large sheets or pieces of timber. Smaller table or bench saws (Fig. 1.12) use a mitre gauge or mitre fence to support the wood when crosscutting. Pieces of timber less than 200 mm long can be dangerous to cut off unless they are held firmly against the mitre gauge (Fig. 1.23). A spacer piece of wood should be cramped to the rip fence, as shown, to prevent a twisting or jamming action occurring. A dangerous kickback could be the result.

Figure 1.23 Use spacer block against fence when crosscutting short pieces; use push stick to clear cut-off pieces past blade

1. Place the mitre gauge in the left-hand table slot and set it at 90° or to the required angle (Fig. 1.17).

2. Check the saw blade for squareness to the table or set it to the required angle if a bevel is required.

3. Set the crosscut blade so that it protrudes 5 or 6 mm above the wood to be cut.

4. Position the blade to cut on the waste side of the marked line (Fig. 1.24). Hold the timber firmly against the mitre gauge and push it smoothly past the blade. Use a push stick to clear the saw for the next cut. Never reach across the saw blade.

5. Pull back the mitre gauge before switching the machine off.

Figure 1.24 Blade should not protrude more than 5 or 6 mm above timber when crosscutting

SAWING MITRES

1. Mark the angle to be cut.

2. Set the mitre gauge at the required angle.

3. Hold the timber firmly against the mitre gauge with the blade cutting on the waste side of the line.

4. Proceed to make the cut by moving the timber smoothly past the blade. Use a push stick to remove the waste if necessary.

5. Return the mitre gauge to its starting position ready for the next cut or before switching off (Fig. 1.25).

Figure 1.25 Mitring jig runs in table grooves; helps cut mitres singly or trims both members at once (guard removed for clarity)

REBATING

Rebating on the circular saw is the process of cutting a square or rectangular recess from the corner of a member (Fig. 1.26).

1. *Mark the rebate on the end of the member.*

2. *Set the rip fence at the required width and the blade to the required depth. Check saw blade for vertical.*

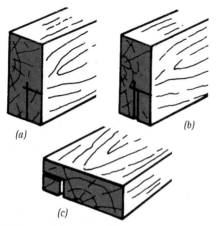

(a)

(b)

(c)

Figure 1.26 Steps in cutting a rebate: (a) Mark rebate; (b) Make first cut; (c) Complete rebate with second cut

3. *Make the first cut. Use a feather board to ensure that the cut remains parallel to the edge.*

4. *Set up for the second cut and complete the rebate.*

CUTTING TENONS

While rebates, tenons, grooves and trenches may be cut more rapidly with the use of a drunken or wobble saw (tapered washers cause the blade to cut a wider kerf) or a dado head (a special wide blade with cutter inserts), a conventional circular saw blade can be used to perform these tasks.

Tenons are machined simply by making multiple cuts across the grain as shown in Figure 1.27. A jig for cutting tenons (Fig. 1.28) may be constructed to facilitate more rapid cutting.

CUTTING MANUFACTURED BOARDS

Manufactured boards such as plywood, particle board and medium density fibre board (MDF) need to be cut using a tungsten tip blade only because the adhesive used to make the boards (urea formaldehyde) is very abrasive. Some table circular saws come with a scribing blade designed especially for the cutting of veneered, melamine or laminate coated boards. This is a small diameter blade located in front of the main blade and is set to cut through the surface of the board before being cut by the main blade, which produces a clean chip-free cut.

If your saw table does not have a scribing blade, a high quality cut can be achieved using a 70 tooth carbide triple-chip saw blade as shown in Figure 1.14. As in cross cutting you need to set depth of cut to 5–6 mm above the cutting surface for the smoothest cut.

A dust mask should be worn at all times when working with MDF or particle board because the urea formaldehyde used in the boards is a health hazard.

Figure 1.27 Set cut-off guide at 90° and adjust rip fence to length of tenon; use multiple cuts to remove waste

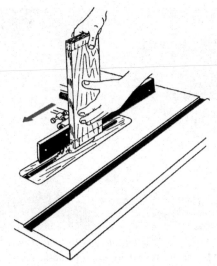

Figure 1.28 Jig for cutting tenons uses rip fence and saw table as guide

SAFETY RULES FOR THE CIRCULAR SAW

1. **General safety.** *Observe the general safety rules given at the beginning of this chapter. The most universal safety rule is to understand how a machine operates and to use it thoughtfully.*

2. **Hearing protection.** *Wear earmuffs to protect your hearing.*

3. **Blade height.** *Adjust the blade so that it only projects slightly (5 mm) above the wood for crosscutting or maximum height for ripping.*

4. **Guard.** *Always use a safety guard if the saw teeth project above the stock being cut.*

5. **Riving knife or splitter attachment.** *Whenever possible (especially when ripping) use the riving knife to prevent kickback.*

6. **Push stick.** *Use a push stick when ripping narrow stock.*

7. **Blade condition.** *Do not use a blunt blade as it tends to burn the wood and jams in the cut.*

8. **Stance.** *Stand to one side of the saw to avoid stock that may kick back. Never reach over the saw blade to support or catch a piece of stock.*

9. **Freehand cutting.** *Avoid sawing freehand as it can be dangerous.*

10. **Wide crosscutting.** *Take special care with wide crosscutting as the stock tends to slew around and catch on the blade. This problem can be avoided by using a shortened fence.*

JOINTERS

The jointer is an electrically driven planer, used mainly to dress the faces and edges of timber. It can also be used to plane bevels, chamfers, tapers and rebates. The size of a jointer is specified by referring to its maximum width of cut, the sizes varying from 100 to 400 mm. The cutter heads revolve at around 3600 rpm.

PARTS AND USES

The main parts of the jointer are shown in Figure 1.29.

Table—*Consists of two parts; the front table which is adjustable for depth of cut and the rear table which is fixed at the same height as the blade (it can be adjusted if required).*

Cutter head—*(Fig. 1.30) The section of the machine that houses the three (sometimes four) knives.*

Fence—*Used to guide the timber and is adjustable for width and angle of cut. The fence should be adjusted to expose only the required amount of blade when dressing a piece of timber.*

Guard—*Covers the cutting knives and swings out of the way when a piece of timber is passed over the knives.*

Table adjustment—*Consists of knobs located under the table. The table is raised and lowered to achieve the depth of cut required.*

Base—*The supporting frame which also houses the motor.*

PLANING FACES, EDGES AND ENDS

Always check the machine before using it to see that the fence is tightened and the depth of cut is correct. When planing a face, the general rule is not to take more than a 3 mm cut in one pass. On extremely hard timber, much less must be taken in a single pass. A

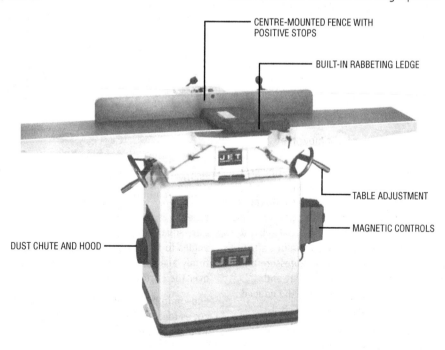

CENTRE-MOUNTED FENCE WITH POSITIVE STOPS

BUILT-IN RABBETING LEDGE

TABLE ADJUSTMENT

MAGNETIC CONTROLS

DUST CHUTE AND HOOD

Figure 1.29 A 150 mm jointer

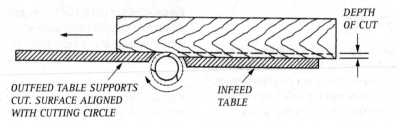

OUTFEED TABLE SUPPORTS
CUT. SURFACE ALIGNED
WITH CUTTING CIRCLE

INFEED
TABLE

DEPTH
OF CUT

Figure 1.30 Blades on cutter head are set to level of rear table

piece of timber can be completely dressed on a jointer if the cutter knives can cut the entire width of the timber in a single pass. After checking the depth of cut, determine the grain direction (Fig. 1.31). The timber should be passed over the cutter so the cutting action goes with the grain, thus producing a smooth surface.

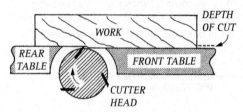

DEPTH
OF CUT

WORK

REAR
TABLE

FRONT TABLE

CUTTER
HEAD

Timber should be passed over cutter head so that fibres tend to lie down, i.e. with the grain

Figure 1.31 Place timber on jointer so that blades are cutting with grain

When dressing a piece of timber that is slightly bowed, either remove a small amount from each end or start dressing the timber from the middle to correct the bow on one face then make a full pass to finish. As the timber is passed over the cutter, transfer the pressure on the wood from the front table to the rear table. As the board is straight after passing over the cutter, putting pressure on the rear table keeps the timber flat and straight. If the timber is warped, the cupped face should be dressed first (Fig. 1.32).

Where possible, use a push stick to apply pressure and guide the timber over the cutters (Fig. 1.33).

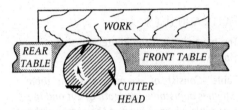

WORK

REAR
TABLE

FRONT TABLE

CUTTER
HEAD

Figure 1.32 Dress cupped face first if timber is warped or bowed

Figure 1.33 Use push stick for safety (guard removed to show cutter head)

When dressing an edge, check the fence with a try square to make certain it is at right angles to the table. Then hold the timber firmly against the fence as it is pushed slowly over the cutter.

End-grain may be dressed using a 1 or 2 mm cut as long as the timber is wider than 200 mm. To prevent the end-grain chipping out at the end of the cut, a chamfer or short pass may be cut on the trailing end first (Fig. 1.34).

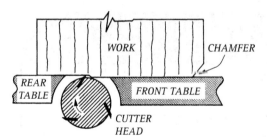

Figure 1.34 Chamfer trailing edge when dressing end-grain

PLANING CHAMFERS AND BEVELS

The jointer can also plane chamfers and bevels if the fence is adjusted to the required angle and the timber held firmly against the fence as it is passed over the cutter. (If the chamfer or bevel is to run on all edges, plane the end-grain first to prevent chipping.)

Note: the angle may be planed with the fence angled in or angled out.

PLANING A REBATE

Most jointers have the facility to plane a rebate if the fence is adjusted so that the amount of blade exposed equals the width of the rebate. The front table should then be lowered to the depth of the rebate—if the depth is greater than 10 mm a number of passes will be required.

OPERATING ADJUSTMENTS

The main adjustment required on the jointer is the setting of the blades in the cutter so they are level and parallel with the rear table. This can be done by loosening the blades and tapping them down or prying them up until they are level. Check the adjustment using a couple of try squares sitting on the rear bed (Fig. 1.36).

Figure 1.36 Check rear table against cutter blades for evenness

Figure 1.35 Dressing a bevel

THICKNESSERS

The thicknesser is also known as a thickness planer. It will plane to a controlled thickness one side of the timber or material at a time.

The thicknesser will not remove twist, wind or warp from a board. The pressure of the infeed and outfeed rollers flattens the board as it passes through, but it springs back on exit.

SAFETY RULES FOR THE JOINTER

1. **General safety.** *Observe the general safety rules given at the beginning of this chapter.*

2. **Guard.** *Make sure the safety guard is in place and ready to use at all times.*

3. **Blades.** *Check to see that blades are sharp and firmly fixed.*

4. **Stance.** *Always stand to the left of the machine to avoid possible kickback.*

5. **Push board.** *Use a push board when surfacing timber. Do not attempt to plane boards less than 300 mm in length. Plane short boards by hand.*

6. **Direction of cut.** *Determine the direction of grain on the timber so that the cut aligns with the grain and leaves a smooth surface.*

7. **Warped timber.** *Always surface the concave side of a board first. Twisted timber may require hand-planing before being dressed on the jointer.*

Therefore, any warp should be first removed from the face that is to be on the table by dressing it on the jointer or with a hand or portable electric plane.

The size of the thicknesser is designated by the maximum width and thickness of material it can handle. For instance, the machine illustrated in Figure 1.37 would be referred to as a 320 × 165 mm thicknesser.

PARTS AND USES

The main parts of a thicknesser are shown in Figure 1.37.

Cutter head—*Consists of three evenly spaced knives or blades and on most machines rotates at 3500 to 5000 rpm.*

Rollers—*Infeed rollers push the timber through the cutter head, while the outfeed rollers hold it down on the table (Fig. 1.38).*

Chip breaker and pressure bar—*Press the workpiece down firmly against the table of the thicknesser.*

Delivery roller—*Keeps the workpiece moving through the machine once its end passes*

through the infeed rollers. It is located at the rear of the pressure bar.

Figure 1.37 A 320 × 165 mm thicknesser or thickness planer

Table—*The flat surface that supports the workpiece.*

Thickness control handwheel—*Adjusts the height of the table and determines the thickness of material remaining after the cut.*

Feed control lever—*Determines the rate of feed of the workpiece through the thicknesser.*

PLANING TO THICKNESS

1. *Plane one face flat on the jointer or with a hand-planer. Place this face downwards on the thicknesser table.*

2. *Set cutter head to take a light cut.*

3. *Determine the grain direction on each face to be surfaced and mark with an arrow. When feeding the workpiece into the thicknesser, the grain on the surface being dressed should point towards the front or infeed side of the machine.*

4. *Select the feed rate required.*

5. *Start the machine and feed the workpiece into the front of the machine.*

6. *Move to the delivery end on the machine and support the workpiece as it passes through (Fig. 1.38, page 22).*

7. *Until required thickness or width is obtained, reverse the faces of the board with each cut to minimise any tendency for it to warp.*

If a number of pieces are to be dressed to the same thickness, each piece should be surfaced in turn, at the one setting. This is to ensure uniform width or thickness of material.

DRILL PRESSES

Drill presses are made in bench or floor models. The only difference between the models is the length of the central column. They are most frequently used for drilling

---

SAFETY RULES FOR THE THICKNESSER

1. **General safety.** *Follow the general safety rules listed at the beginning of this chapter.*

2. **Blade protection.** *Check the material to be planed for the presence of nails or any objects that could damage the blades.*

3. **Grain direction.** *Place the timber to be planed on the table of the thicknesser so that the grain direction on the top of the board points towards you.*

4. **Board size.** *Do not attempt to surface boards less than 300 mm in length as delivery rollers may not be able to feed the piece through.*

5. **Stance.** *Always stand to one side of the machine while it is in operation to avoid kickback.*

6. **Depth of cut.** *Restrict the maximum cut to 3 mm as a deeper cut will overload the machine.*

7. **Push board.** *Keep fingers away from the table when the machine is in operation. Use a push board if it is necessary to move a stalled piece past the delivery rollers.*

8. **Hearing protection.** *Wear earmuffs or earplugs to avoid hearing damage.*

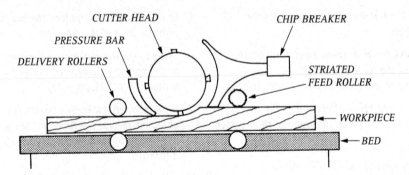

CUTTER HEAD

CHIP BREAKER

PRESSURE BAR

DELIVERY ROLLERS

STRIATED
FEED ROLLER

WORKPIECE

BED

Figure 1.38 Cross-sectional view of thicknesser

or boring holes,[1] but attachments are available to carry out a number of other woodworking operations, such as sanding, morticing and routing (Fig. 1.39).

The size of a drill press is determined by measuring twice the distance from the centre of the chuck to the front of the vertical column. Some manufacturers also specify the maximum travel of the spindle in determining the capacity of the machine. Thus a 400 mm drill press with a 150 mm stroke is one that measures 200 mm from chuck centre to column and has a spindle that can travel 150 mm. The 400 mm designation indicates that the machine will drill a hole in the centre of a 400 mm circle.

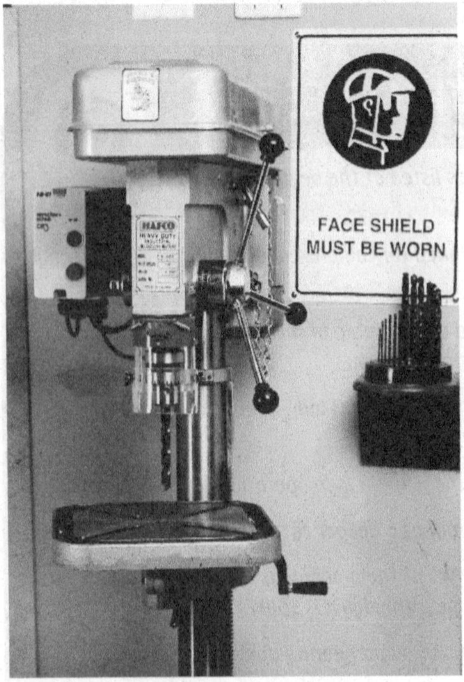

Spindle speeds vary between 300 and 3000 rpm. The drive is mostly by V belt on four- or five-step cone pulleys. The correct speed for drilling depends upon the material being drilled and the diameter of the bit or cutter being used.

PARTS AND USES

Base—*A heavy cast-iron machined casting with a horizontal face. It can be bolted firmly to the bench or floor. The base is usually slotted similar to the table so that it extends the capacity of the machine.*

Column—*A machined steel tube that fits into the base and supports the table and head.*

Table—*Clamped to the column and can be moved vertically or rotated to suit workpieces of various sizes. The table on most drill presses can be tilted for angle boring. A rack and pinion mechanism for raising and lowering the table is available for most machines also.*

FACE SHIELD MUST BE WORN

Figure 1.39 A bench drill press

[1] Drilling and boring mean the same thing in woodworking.

Head—*The top assembly which includes the spindle, pulleys, bearings and motor.*

Chuck—*Most drill presses use a 'Jacobs'-type geared chuck with a 0 to 13 mm capacity. Its purpose is to hold securely the bit or cutter. It is adjusted by means of a geared key.*

Feed lever—*Raises or lowers the spindle during drilling.*

Depth stop—*A bolt with locking nuts attached to the quill to control the depth of drill. Some machines use a revolving, calibrated stop.*

ADJUSTING THE DRILL-PRESS TABLE

Most drill-press operations require the table to be set at 90° to the spindle. Check this with a try square as shown in Figure 1.40. An adjustment nut is situated under the table. A sliding bevel may be used in the same fashion as the try square to set the tilt of the table for other angles. Some machines have locating pins to positively set the table at 45° and 90°. The table is either raised or lowered by hand or with an elevating mechanism. A locking lever fixes it firmly to the column.

DRILLING HOLES

1. Lay out the position of the hole with crossed pencil marks (deviation from centre is immediately visible). A bradawl hole or centre punch mark acts as a pilot and helps to guide the drill.

2. Select the correct bit or accessory and install it firmly in the chuck.

3. Adjust the speed of the spindle to suit the operation to be performed. The speeds for routing should always be as high as possible; however, use lower speeds for large diameter work.

4. Place the workpiece on the table and cramp in place if necessary (Fig. 1.41). Always use a piece of scrap wood under the workpiece. This stops the drill chipping

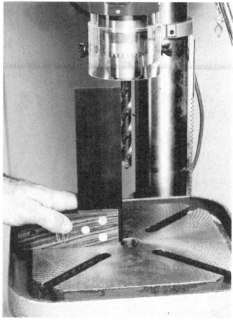

Figure 1.40 Check squareness of drill table and bit

under the job and helps protect the table from damage.

5. Adjust the table to the correct height and tilt angle if required.

6. Adjust the stop to the required depth.

7. Turn the switch on.

8. Hold the job firmly (if not cramped). Apply an even pressure while slowly feeding the bit through the workpiece.

MORTICING

The simplest method of cutting a mortice on the drill press is shown in Figure 1.42. A fence is fixed to the table and, after setting out the mortice on the timber, a series of holes are drilled using a standard twist drill. The holes are drilled closely together and then the drill is used to rout out the remaining waste. The corners of the mortice are squared with a morticing chisel.

Figure 1.41 Securely cramp workpiece to drill press table; waste piece under job helps to produce clean hole

Figure 1.42 Use drill bit to remove waste from mortice

A special morticing attachment is available to be used with a hollow chisel morticing bit (Fig. 1.43). Basically this unit allows the operator to cut a series of square holes in the timber to remove the waste from a mortice. Normally the hollow chisel morticing bit is selected to suit the width of the mortice being cut. The

square holes should be drilled in two stages as shown in Figure 1.44.

Figure 1.43 Morticing attachment with hollow chisel bit

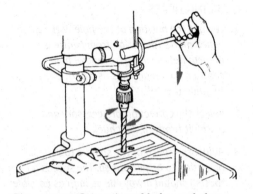

Figure 1.44 Cut series of holes and then remove waste with second series

SANDING

Sanding drums of various sizes can be used on the drill press (Fig. 1.45). These are especially useful for sanding curved work inside cuts and holes as well as outside edges. A custom-made sanding table with an assortment of holes to suit different size drums is an asset. It allows the table to be moved up and down to make more use of the drum.

Figure 1.45 Drum sanding on drill press gives square edge

ROUTING

Most portable electric routers have a spindle speed of 25 000 rpm whereas the top speed of drill presses is usually only 5000 rpm. Therefore, the finish provided by the drill press is not as smooth as that expected of the standard router.

While standard router bits can be used, there are several types of bits designed to be used at lower rpm—such as the drum and plug cutter shown in Figure 1.45.

Material is always fed from left to right. This means that the work must be moved in against the rotation of the cutter. The thrust of the cutter should cause the piece to be pressed against the fence.

The depth of cut should not exceed 3 mm per cut, or pass. If a deeper cut is required, a number of passes should be made with the bit set deeper each time. Small cutters only are recommended because most drill presses will not withstand a side load. Attach the fence

Figure 1.46 Routing on the drill press

firmly to the table (Fig. 1.46) and control the depth of cut by elevating the drill table.

Routing on the drill press is to be avoided where possible as the exposed cutter is extremely dangerous.

BORING OR CUTTING LARGE HOLES

There are several drill press attachments available for boring or cutting large holes.

Forstner bit—*Mostly used for boring shallow holes to remove waste cleanly with very little centre point penetration. The outer edge of this bit helps to control its cutting action so that it can be used to cut screw pockets (Fig. 1.47) or overlapping holes.*

Sawtooth bit—*Used for boring larger holes in harder woods where a quicker cutting action is required (Fig. 1.48).*

Holesaw—*Cuts a hole with minimum effort and waste. An added bonus is that it can also produce wheels or circular discs (Fig. 1.49).*

Adjustable wheel cutter—*Cuts holes or discs of various diameters.*

The wheel cutter and holesaw should be used from both sides of the workpiece to ensure a clean cut (Fig. 1.50).

Figure 1.47 Forstner bit is used to cut screw pocket, sloping fence helps to produce uniform pockets

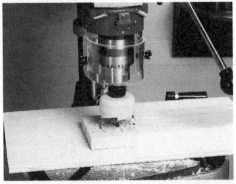

Figure 1.49 Holesaw is used to bore large shallow holes

Figure 1.48 Sawtooth bit will give a cleaner hole if used from both sides

Figure 1.50 Adjustable wheel cutter, cramp job securely and rotate attachment by hand before switching on

SAFETY RULES FOR THE DRILL PRESS

1. **General safety.** *Observe the general safety rules given at the beginning of the chapter.*

2. **Bits and cutters.** *Use the correct bits or cutters for the job and make sure that they are securely fixed in the chuck. Remove the chuck key immediately.*

3. **Speed.** *Check and adjust the speed of the spindle to suit the job in hand. Use the recommended speeds for different materials and bit sizes.*

4. **Securing workpiece.** *Clamp small workpieces firmly to the drill table or in a machine vice. Larger pieces may be hand-held, but positioned to swing away from you if the cutter catches.*

5. **Attachments.** *Morticing, routing and shaping attachments must be properly fastened and adjusted.*

LINISHERS AND DISC SANDERS

The linisher is a type of belt sander, used to sand both flat and curved—convex and concave—surfaces, whereas the disc sander is restricted to flat surfaces and convex curves.

The size of the linisher is usually designated by the width of the belt. The size of the disc sander goes by its diameter. In combination machines, both the disc and the belt are driven by a single motor via pulleys and belts.

There is more information on sanders in Chapter 7.

PARTS AND USES

Abrasive disc—*Consists of a sheet of abrasive paper on a metal disc. A special non-hardening contact adhesive, called disc cement, is used to attach the abrasive paper.*

Table—*Can be tilted to sand bevelled edges. Check the angle of tilt with a try square (Fig. 1.52).*

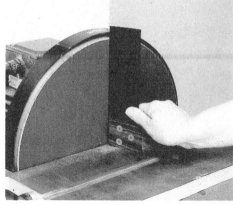

Figure 1.52 Use a try square to check angle of table to disc

Figure 1.53 Tracking adjustment knob controls position of belt on drums

Mitre gauge—*Holds stock at various angles. Most disc sander tables are provided with a slot to guide a mitre gauge.*

Belt—*Consists of an abrasive-coated cloth travelling over two drums and a platen. One drum is fixed in position and driven by a belt connected to the motor. The other drum can be moved to tension the belt or make it track properly (Fig. 1.53). Belt speed on some machines is variable but on fixed belt sanders it is around 1000 m/min.*

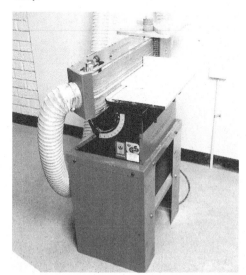

Figure 1.51 A linisher or combination drum and belt sander

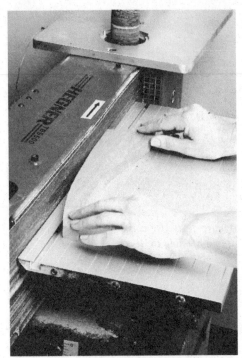

Figure 1.54 Outside curves are sanded square to face using fence as guide

Fence—*Can be set to sand square or to bevel (Fig. 1.54).*

BASIC OPERATIONS ON THE LINISHER

Disc sanding

The disc sander is particularly suited to end-grain sanding (Fig. 1.55). Squaring ends is best achieved by using the mitre gauge set at 90°. Rounding corners, bevelling and chamfering operations can also be carried out at a variety of angles by tilting the table.

Belt sanding

The belt sander is more suitable for larger pieces than is the disc sander. Faces can be sanded along the grain if the material is thick enough to allow fingers to grip safely. Oversized pieces can be sanded by removing the stop or table at the end of the belt sander. Concave surfaces can be sanded on the drum when the guard is removed (Fig. 1.56).

Figure 1.55 Sanding end-grain with a disc sander

Figure 1.56 Use drum section of belt sander to sand concave surfaces

SAFETY RULES FOR THE LINISHER AND DISC SANDER

1. **General safety.** *Observe the general safety rules given at the beginning of this chapter.*

2. **Hand protection.** *Keep fingers at least 50 mm away from discs or belts at all times. Do not attempt to sand small objects that could get caught in clearance spaces on the machine.*

3. **Table clearance.** *The distance between the disc and the table should be less than 2 mm.*

4. **Workpiece positioning.** *Only use the upper downward-moving quadrant of the disc for sanding. The other side of the disc will tend to throw the job.*

5. **Abrasive belts and discs.** *Select the correct grade of grit and size of belt or disc for the sanding operation to be done.*

6. **Belt tracking.** *Place the belt on the machine so that the arrow on the belt is facing the direction of movement and adjust its position for running.*

7. **Extraction.** *Machines should only be used when the dust extractor is switched on.*

QUESTIONS

?

1. *Describe how to make a bandsaw blade track correctly and how to adjust the tension.*
2. *Why is it important not to attempt to cut round stock on the bandsaw?*
3. *What is the purpose of a tilting arbor on a circular saw?*
4. *Name the three types of blades used on circular saws and sketch the basic tooth shape of each.*
5. *What is the purpose of the riving knife on the circular saw?*
6. *What is the purpose of the movable table on the jointer?*
7. *The main adjustment on the jointer is the setting of the blades. Outline the principal points to keep in mind when setting them back in position after having sent them out to be reground.*
8. *How is the size of the thicknesser designated?*
9. *Describe the procedure to be followed if a number of pieces are to be dressed to the same dimensions on the thicknesser.*
10. *How is the size of a drill press designated?*

SURFING THE NET

Woodworking tools

http://www.vicnet.net.au

GENERAL SAFETY

In the main, power tools have been designed to be efficient and safe to use when proper precautions are observed. Safety rules specific to individual machines are included in the relevant sections throughout the chapter. The following general safety rules apply to all machines throughout the chapter.

1. **Permission.** *Obtain permission from the teacher or instructor before operating any of the machines.*

2. **Eye protection.** *Always wear safety glasses or eyeshields when operating any power machine.*

3. **Hearing protection.** *Always wear earmuffs or earplugs to avoid hearing damage when working on noisy machines.*

4. **Jewellery.** *Metal rings on your fingers can be dangerous if splinters are present in the timber. Remove rings before using any of the machines.*

5. **Clothing.** *Do not wear loose clothing; remove your tie, roll your sleeves up and tie long hair back.*

6. **Dust.** *Protect yourself from the long-term health problems that can result from dust inhalation. Wear a dust mask if an efficient dust extraction system is not available.*

7. **Lighting.** *Make sure that shadowproof lighting is available to provide an unobstructed view.*

8. **Attachments and accessories.** *The instruction manual that comes with each machine often suggests special set-ups and jigs. Familiarise yourself with these.*

9. **Electrical safety.** *Portable or in-built core balance earth leakage circuit-breakers should be included in all workshops (Fig. 2.1). Even though electric supply circuits are protected by a fuse or circuit breaker device, the only completely safe system is the earth leakage core balance relay. They cut off the power before it can harm the operator if the earthing system of a tool fails. Make sure the earth lead on three-pronged electric plugs is properly grounded. Most portable power tools are double-insulated for added safety. These tools must not be earthed, even if supplied with a three-pin plug. Double-insulated tools rely on two layers of internal insulation between the current-carrying parts and all exposed metal sections of the tool.*

10. **Electric cord.** *Take care to arrange the cord so that it is clear of the path of action. A good practice is to hang the cord over one shoulder. Never use the cord while it is coiled as the induced current can cause it to overheat.*

11. **Adjustments.** *Always switch off the power and disconnect the plug from the power outlet before making adjustments to a machine.*

12. **Stopping.** *After switching off a portable power tool, always hold it in your hands until it has stopped. Place the tool in a safe place or return it to its proper place when not in use.*

Chapter 2

Portable Power Tools and Processes

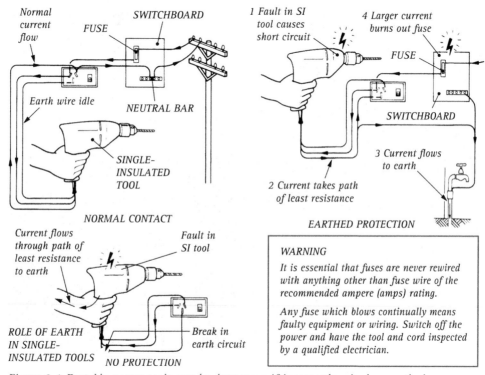

Figure 2.1 Portable power tools can be dangerous if incorrectly wired or earthed.

Labels in figure:

Normal current flow
FUSE
SWITCHBOARD
Earth wire idle
NEUTRAL BAR
SINGLE-INSULATED TOOL
NORMAL CONTACT

1 Fault in SI tool causes short circuit
4 Larger current burns out fuse
FUSE
SWITCHBOARD
3 Current flows to earth
2 Current takes path of least resistance
EARTHED PROTECTION

Current flows through path of least resistance to earth
Fault in SI tool
ROLE OF EARTH IN SINGLE-INSULATED TOOLS
Break in earth circuit
NO PROTECTION

WARNING

It is essential that fuses are never rewired with anything other than fuse wire of the recommended ampere (amps) rating.

Any fuse which blows continually means faulty equipment or wiring. Switch off the power and have the tool and cord inspected by a qualified electrician.

PORTABLE HAND DRILLS

The portable power drill is manufactured in a wide range of sizes, powers and speeds (Fig. 2.2)

TYPES AND SIZES

There are three basic types of portable hand drills:

1. *The pistol-grip is designed for smaller work and one-handed operation (Fig. 2.3).*

2. *The D-handle drill allows the operator to drill more accurately because pressure is applied directly in line with the axis of the drill (Fig. 2.4).*

3. *The spade-handle drill is essentially a heavy-duty machine with greater power and larger chuck capacity than the other two types of drills. It usually operates at a slower speed (Fig. 2.5).*

The size of a portable power drill is determined by its maximum chuck

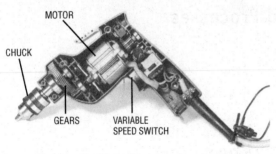

Figure 2.2 *A sectioned hand drill*

Figure 2.3 *A pistol-grip hand drill, designed for smaller work.*

Figure 2.4 *A D-handle hand drill, used for more accurate drilling*

Figure 2.5 *A spade-handle hand drill, used for heavy-duty drilling operations*

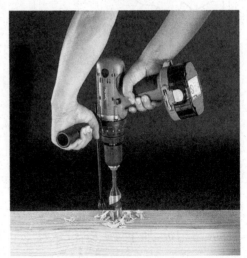

Figure 2.6 *Battery powered drills now have a range of torque settings.*

capacity. Sizes vary from 6 to 25 mm. The drill speed increases as the chuck size decreases as, in order to cut efficiently, smaller drills should run at a higher speed.

Many of the latest drills have forward and reverse switches and variable speed and hammer drilling capabilities. There are also a wide range of accessories available; for

example drill stands, circular saws, hole saws, various bits, sanding discs and drums, and rotary files.

Most modern power drills are double-insulated and relatively safe to use, but great care should be exercised with older types of drills to ensure they are correctly grounded.

Battery powered drills are now available with greater voltage and torque. Keyless chucks have been improved and are more convenient.

DRILLING HOLES

1. *Mark the position of the hole in pencil and indent the wood with a centre punch, bradawl or the point of a nail so that the drill will not wander (Fig. 2.7).*

2. *Select the correct type and size of bit. Fix it firmly in the chuck.*

3. *Position the point of the drill in the mark before switching on the power.*

4. *Guide the drill as accurately as possible (Fig. 2.8). Only apply the pressure necessary to keep the drill cutting in the wood.*

5. *Use a backing piece of wood to prevent the drill from breaking out the fibres if the hole is to be drilled right through.*

6. *Always withdraw the drill from the job before turning it off.*

Figure 2.7 Mark position for hole to be drilled

Figure 2.8 Guide drill carefully, applying just enough pressure to keep it cutting

SAFETY RULES FOR THE PORTABLE DRILL

1. **General safety.** *Observe the general safety rules given at the beginning of this chapter.*

2. **Beware.** *Check for electricity or water lines before attempting to drill in building situations.*

3. **Securing workpiece.** *Make sure the workpiece is securely held when drilling and suitable backing is provided where necessary.*

4. **Chuck key.** *Always remove the chuck key from the chuck immediately after use.*

PORTABLE BELT, FINISHING AND DISC SANDERS

The use of abrasives on timber is very important if a project is to be correctly finished. Hand-sanding with a cork block gives an excellent finish but requires a lot of effort and time. The portable power sander achieves the same finish and requires much less exertion.

Three types of portable power sanders are available, and all are almost essential if a professional finish is required.

PORTABLE BELT SANDERS

Portable belt sanders are used for sanding large surfaces that cannot be easily sanded on a stationary unit (Fig. 2.9).

Figure 2.9 A portable electric belt sander; note use of vacuum hose to remove dust.

An important part of the sander is the cloth abrasive belt (75 and 100 mm are the most common sizes) which runs over two pulleys or drums. The rear pulley, usually rubber-covered and fixed in position, delivers power to the sanding belt from the motor. The front pulley is crowned, tiltable and can be moved forward and backward. The tilting capability allows the belt to be tracked and the longitudinal movement is used to maintain the tension.

Most of the portable belt sanders available are fitted with a built-in fan or impeller and a cloth bag to minimise airborne dust.

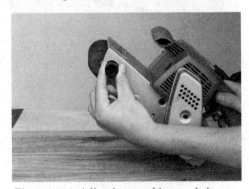

Figure 2.10 Adjusting tracking on belt sander

Using the portable belt sander

Always make sure the power switch is off before plugging in the machine, otherwise, the machine will begin to run and could fall off the bench. Check that the belt is tracking before attempting to use the machine. Raise the sander above the job before switching on and lift it off the job before switching off.

Figure 2.11 Fitting a belt

Make sure the workpiece is secured to the bench in a position that facilitates sanding with the grain. The belt sander is used with the grain when sanding faces and edges. Depending on the state of the surface to be sanded, it may be necessary to start with a coarse (No. 60) grit, switch to a medium (No. 80) grit and finish with a fine (No. 100) grit. At this point it is advisable to use the straight line finishing sander. Some handsanding with a sanding block may also be necessary.

Refer to the belt sander travel diagram (Fig. 2.12(b)) for the method of sanding a large surface. Pressure should not be applied to a belt sander and it should always be kept moving, or it will give an uneven surface. Some manufacturers provide a special sanding shoe, which is actually a narrow surrounding guide for extra-smooth flat sanding.

PORTABLE FINISHING SANDERS

Finishing sanders are designed for final finishing and are either orbital or oscillating in motion. The orbiting action tends to leave circular scratches on a wood surface so a better finishing sander has a straight line reciprocating action.

Combination finishing sanders are available with a switch allowing the selection of either straight line or orbital action.

(a)

SIDE VIEW

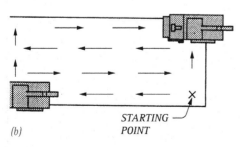

STARTING — POINT

(b)

Figure 2.12 (a) Switch sander on before lowering it onto job, (b) Method of using belt sander on large surface

Figure 2.13 A portable finishing or orbital sander

Figure 2.14 A small hand-finishing sander with dust bag, used to get into tight spots

The uppermost part of the sander consists of a motor housing and a handle assembly. The switch is usually built into the handle housing. The base of the sander consists of a suspended pad and a clamping system which permits abrasive paper to be fastened to the pad.

PORTABLE DISC SANDERS

Disc sanders with flexible discs are limited in their use in the woodshop because of the circular scratches they often leave. The orbiting disc sander overcomes this problem if a suitable grade grit is used. These sanders are particularly suited for sanding curved surfaces, such as boat hulls.

The sanding discs are usually attached to the pad by means of a self-adhesive backing or Velcro.

Note: Dust particles from a sander are very dangerous if inhaled. The dust contains not only the abraded timber but also a significant amount of the abrasive material. Most good quality sanders come with a dust bag which helps remove a lot of the dust, but the better-designed machines are equipped with an extraction system such as the unit shown in Figure 2.15.

Figure 2.15 A Makita dust extraction system can be connected to a number of different portable power tools.

SAFETY RULES FOR THE PORTABLE SANDER

1. **General safety.** *Observe the general safety rules given at the beginning of this chapter.*

2. **Abrasive belts.** *Select a suitable grade grit for the job and check for correct belt or disc.*

3. **Belt direction.** *Install the belt so that the arrow on the inside of the belt is pointing in the direction of rotation.*

4. **Belt tracking.** *Make sure the belt is tracking correctly before using the sander. Adjust if necessary.*

5. **Plugging in.** *Always hold the machine with one hand when plugging the lead into a power point as the sander could 'walk' off the bench.*

6. **Starting.** *Raise the sander above the workpiece before switching on and allow it to reach full speed before proceeding.*

7. **Hand protection.** *Keep both hands on the handles while operating the sander.*

8. **Stopping.** *Place the sander on its side when not in use.*

PORTABLE JIGSAWS

The portable electric jigsaw is a versatile machine designed to cut a wide variety of materials, such as wood, metal, plywood, particle board, custom board, plastic and leather.

There is little variation in the sizes of jigsaws, as they are all designed for one-handed use in most operations. Any variation is in the robustness and power of the machine. Most of the jigsaws available are double-insulated, and have a single-phase motor with a 300 to 450 watt rating and 3100 blade strokes per minute. The heavy-duty machine with a 25 mm stroke will readily cut through 50 mm thickness of timber and will cut both curved and straight lines.

Most of the better quality jigsaws have an orbital and reciprocating blade motion. The blade cuts on the up stroke and backs away on the down stroke. This action reduces drag on the down stroke and helps the blade run cooler.

Figure 2.16 A variable speed jigsaw.

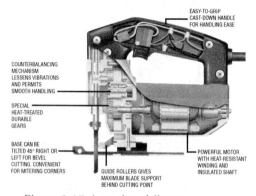

EASY-TO-GRIP CAST-DOWN HANDLE FOR HANDLING EASE

COUNTERBALANCING MECHANISM LESSENS VIBRATIONS AND PERMITS SMOOTH HANDLING

SPECIAL HEAT-TREATED DURABLE GEARS

BASE CAN BE TILTED 45° RIGHT OR LEFT FOR BEVEL CUTTING. CONVENIENT FOR MITERING CORNERS

GUIDE ROLLERS GIVES MAXIMUM BLADE SUPPORT BEHIND CUTTING POINT

POWERFUL MOTOR WITH HEAT-RESISTANT WINDING AND INSULATED SHAFT

Figure 2.17 A sectioned jigsaw.

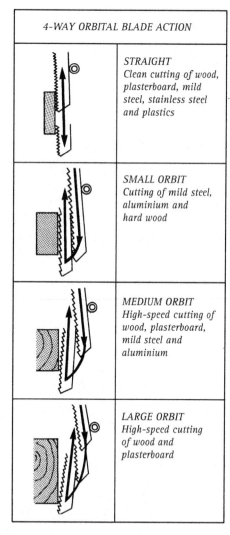

4-WAY ORBITAL BLADE ACTION	
	STRAIGHT Clean cutting of wood, plasterboard, mild steel, stainless steel and plastics
	SMALL ORBIT Cutting of mild steel, aluminium and hard wood
	MEDIUM ORBIT High-speed cutting of wood, plasterboard, mild steel and aluminium
	LARGE ORBIT High-speed cutting of wood and plasterboard

Figure 2.18 Blade stroke, on some jigsaws, can be altered to suit the thickness of material.

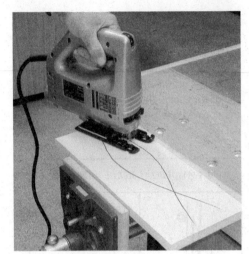

Figure 2.19 Hold timber securely in vice when freehand cutting; do not withdraw blade from cut until it has stopped moving.

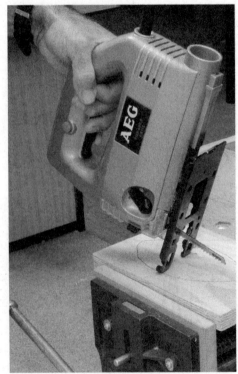

Figure 2.20 Making a plunge cut. Rest base of jigsaw firmly on job before commencing; hold machine with both hands

Figure 2.21 Pivot base on job as shown until blade cuts or plunges through.

Figure 2.22 Some form of guide, such as the one shown above, is recommended for crosscutting

JIGSAW BLADES

Manufacturers vary in their methods of fixing the blade to the saw and there are different blades available for cutting wood, plastic, metal and so on. The long blades are designed for cutting thick wood (50 mm) and the short blades are for cutting small radius curves and intricate designs. It is important to select the right type of blade for the material being used and to set the correct cutting angle for the table, which may be set from 45° to 90°.

Study the operator's manual thoroughly before using the machine as there are instructions for adjustment specific to each brand.

FREEHAND CUTTING AND USING GUIDES

Most freehand cutting is to curved lines whereas most straight cutting along and across the grain (ripping and crosscutting) is done using the attached guides. A guide for cutting circles is normally included with each machine as standard equipment.

SAFETY RULES FOR THE PORTABLE JIGSAW

1. **General safety.** *Observe the general safety rules given at the beginning of this chapter.*

2. **Blade selection.** *Select the correct type and size of blade to suit the machine being used and the job in hand.*

3. **Securing workpiece.** *The reciprocating action of the blade requires the job to be securely held at all times to avoid blade breakage. Support the workpiece, especially at the end of the cut.*

4. **Starting the cut.** *Always keep the blade clear of the job when starting the cut.*

5. **Stopping the cut.** *Always wait until the blade has stopped moving before removing it from the cut.*

PORTABLE CIRCULAR SAWS

Most portable circular saws consist of a motor mounted in a housing and a circular blade driven through a gear train (Fig. 2.23).

The size of the portable circular saw is designated by the largest blade that can be fitted to the machine. The sizes on the market range from 110 to 415 mm. The most common saws have spur or bevel gears that tend to be noisy in operation. The hypoid circular saw is different in that the motor shaft is parallel to the blade which is driven through a hypoid right-angle gear. The advantages of this

machine are that it runs more quietly because of the hypoid gearing and the position of the handle gives better control.

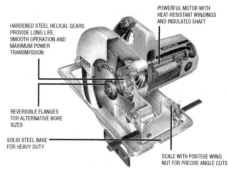

Figure 2.23 A sectional view of a portable circular saw

Figure 2.24 Battery powered circular saws are excellent for light work in the woodshop.

Figure 2.26 Use a box spanner to fix blade onto spindle

Figure 2.25 Setting depth of cut; make sure all adjustments are firmly fixed before using saw

Figure 2.27 Position spring-loaded guard as shown when cutting thin boards

The power of the motor varies between 850 and 1850 watts according to the size of the saw. The speed of the blade also varies according to its diameter, ranging between 3500 and 4500 rpm under load.

OPERATING ADJUSTMENTS

A fence is provided for ripping or making parallel cuts and this is simply adjusted to the required width. The depth of cut of the blade is adjusted by moving the base or table up or down as shown in Figure 2.25. The correct depth of cut for ripping or crosscutting should equal the thickness of the material plus 4 or 5 mm. Portable circular saws are designed to cut at angles between 45° and 90°. The base or table is pivoted and a calibrated quadrant allows the blade to be fixed at the required

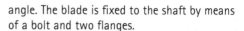

angle. The blade is fixed to the shaft by means of a bolt and two flanges.

USING THE CIRCULAR SAW

Place the base of the machine on the timber so that the blade is just clear. Use some form of guide, if possible, such as the fence straight edge. Freehand cutting should not be attempted unless there is a line to work from. Switch the motor on and wait until the blade has reached its full speed before starting the cut.

On the completion of the cut, release the switch. The guard should automatically spring back and cover the blade but do not put the saw down until it has completely stopped.

Figure 2.28 Making a bevel cut; attach a guide strip to help control cut

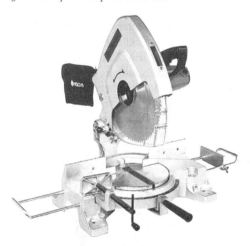

Figure 2.29 Hitachi mitre saw is used to produce accurate square and angled joints on relatively narrow timber

Figure 2.30 Makita slide compound saw— serves many functions of radial arm saw and provides much greater capacity than a mitre saw, but also serves a similar purpose

SAFETY RULES FOR THE PORTABLE CIRCULAR SAW

1. **General safety.** *Observe the general safety rules given at the beginning of this chapter.*
2. **Blade selection.** *Do not use a blunt blade or one that has lost its set.*

3. **Power.** *Always disconnect the machine from the power when making adjustments.*

4. **Starting.** *Rest the toe of the base on the job before switching on and do not commence cutting until the motor has reached full speed.*

5. **Securing workpiece.** *The workpiece should be securely held and the power lead should be positioned clear of the saw's path.*

6. **Hearing protection.** *Always wear earmuffs or earplugs to avoid hearing damage.*

7. **Holding.** *Use both hands to guide the circular saw to avoid kickback.*

8. **Freehand cutting.** *Avoid freehand cutting where possible kickbacks can occur.*

9. **Cutting depth.** *Always check for the correct depth adjustment before making a cut.*

10. **Bevel cutting.** *Check for the correct bevel adjustment before making a cut.*

11. **Ripping or crosscutting.** *Use the fence for ripping along the grain and the mitre guide for crosscutting.*

PORTABLE ELECTRIC ROUTERS

The portable electric router is a precision power tool with a wide range of applications in the woodwork shop. Once set up, the router facilitates the fast and accurate production of intricate joints, decorative cuts and shaped edges. It can be used freehand or with the aid of guides or templates.

Routers vary in size and power depending on the types of operations for which they are designed (Fig. 2.31). A router is named by the size of its motor and the chuck size. A small capacity router with a 6 mm collet chuck and a 440 watt motor would be used for fine work such as edge trimming, whereas at the top end of the scale, a router with a 1850 watt motor and a 12 mm collet chuck would be used for large cuts (Fig. 2.32). The speed of a router may also vary with its capacity; for example, some edge trimmers rotate at 30 000 rpm and others at a maximum of 18 000 rpm, but the average 1500 watt router rotates at about 23 000 rpm.

Figure 2.32 Plunge router allows operator to start cut at any stage on work

(a)   (b)   (c)

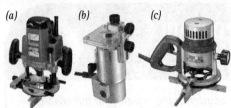

Figure 2.31 Three different types of routers: (a) plunge; (b) edge trimmer, (c) fixed base

The newer electronic routers have a speed control which varies the speed from 8000 to 18 000 rpm and holds the cutting speed while the cut is being taken.

The plunge router (Fig. 2.32) allows the cutter to be lowered into the timber in a plunge-type cut. The advantage of the plunge router is evident in joinery template operations where lots of material has to be removed. These include trenches, housings, tenons, mortices and template work. A cut can be started in the middle of a piece of timber, whereas a fixed base machine has to be tilted to ease the bit into the cut, often resulting in loss of control and unsafe routing.

FIXED BASE ROUTER

The standard fixed base router is suitable for general routing operations where the depth of cut is fixed and the base locked in place (Fig. 2.33).

Figure 2.33 A D-handle, fixed base router.

Routers should always be used against the rotation of the cutter. That is from left to right on the edge facing the operator.

PARTS AND USES

Chuck—*Routers use a collet chuck, which tightens down onto a parallel shank. The diameter of the shank varies. A cutter must fit exactly into the collet. Be careful when selecting the correct size cutter for the collet as there are two standards of measurement used— imperial and metric. An equivalent metric cutter should not be used in an imperial collet as a 6 mm cutter shank is smaller in diameter than a ¼ inch (6.35 mm) shank. The cutter may work loose causing an accident. The most common imperial collets sizes are ¼, ⅜ and ½ inch diameter. The most common metric collet sizes are 6 mm and 12 mm. There are two main types of collet chucks for the router, the split collet and the fingered collet, the fingered giving the better holding power.*

Base—*The surface in contact with the wood. This is the surface which the router rides and therefore it controls the depth of cut.*

Depth adjustment—*Depends on the type of router. If it is a fixed base router, the depth of cut adjustment is done with a threaded ring running on the body of the router (Fig. 2.33). On the plunge router, a post slides up and down and measures the depth of cut. The base may be fixed in place by a locking lever and a threaded post.*

Handles—*The two handles are conveniently placed for easy grasp and control. There is the D-type handle as on the router shown in Figure 2.33 which has the switch in the handle, or the normal two-knob handle as on the router shown in Figure 2.32.*

Bits—*The various cutters fitted into the collet chuck. In recent years there have been many developments in the technology of router bit design. Overall performance has been improved by the use of new materials, coatings and edge tippings, along with the refinement of geometric cutting designs.*

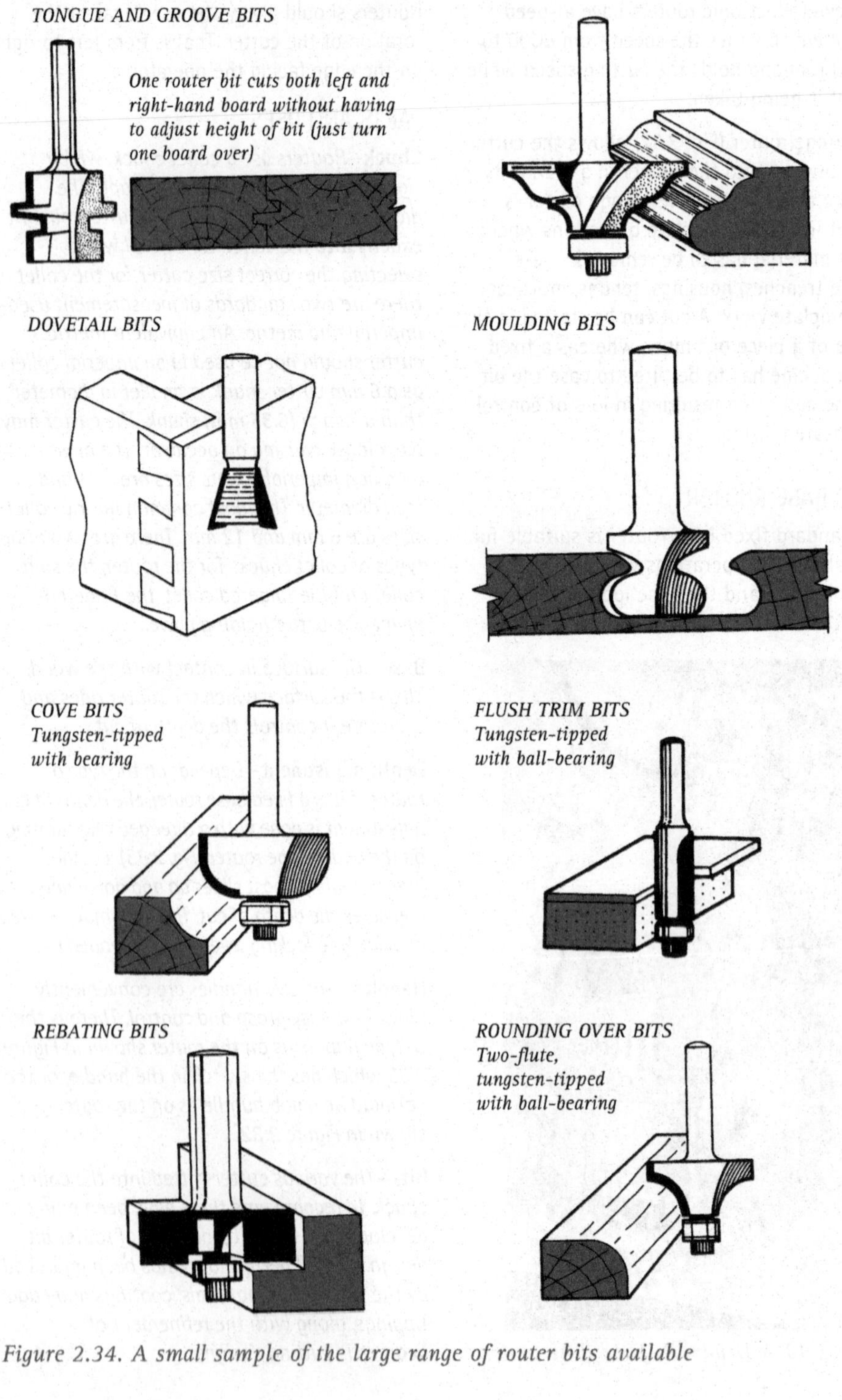

TONGUE AND GROOVE BITS

One router bit cuts both left and right-hand board without having to adjust height of bit (just turn one board over)

DOVETAIL BITS

MOULDING BITS

COVE BITS
Tungsten-tipped
with bearing

FLUSH TRIM BITS
Tungsten-tipped
with ball-bearing

REBATING BITS

ROUNDING OVER BITS
Two-flute,
tungsten-tipped
with ball-bearing

Figure 2.34. A small sample of the large range of router bits available

ROUTER BITS

Most router bits in use today are made of high-speed steel or are tungsten carbide tipped. High technology materials are used to make industrial bits—the polycrystalline diamond bit was developed for hard use on particle board (and costs fifteen times the price of a carbide bit) and the ceramic bit was developed to cut the more abrasive materials such as medium density fibreboard and particle board. Research has shown that the ceramic bit will last six times longer than the carbide bit, yet only costs 50% more.

There are router bits available now that will achieve just about any cut desired (Fig. 2.34). One-piece bits are used to cut trenches, hollows and shape edges. Shaper cutters can be assembled to cut custom profiles and cutters with roller-bearings are available for following edges.

SETTING UP THE ROUTER

1. *Select the router bit and fit it into the collet chuck (Fig. 2.35). (Make sure it is the correct-size chuck for the bit.) The bit should be inserted fully, then pulled out approximately 2 mm.*

Figure 2.35 Fitting router bit into chuck; check shank size—do not mix imperial and metric

2. *Tighten the collet chuck but do not overtighten as this makes it difficult to remove the bit.*

3. *Set the depth of cut. Some plunge routers have a vernier-type adjustment which, once zeroed in, allows the cutter to be accurately adjusted for the depth of cut. On the fixed base routers, it is necessary to measure the depth using a rule (Fig. 2.36).*

4. *Make sure that the base is securely tightened.*

Figure 2.36 Use straight edge to check position of bit relative to base of router

ROUTING GROOVES, TRENCHES, REBATES AND EDGES

The edge-guide attachment supplied with the router aids the cutting of grooves and trenches and rebates parallel to the edge or end of a piece of timber. Edge shaping with a piloted router bit is one of the simplest routing operations. This bit follows a straight or curved edge and shapes it uniformly. The edge must be true to shape and smooth because the bit reproduces any irregularity it passes over.

1. *Select a suitable cutter and fasten it in the chuck.*

2. *Adjust to the correct depth of cut.*

Figure 2.37 Using a piloted router bit to run a bullnose moulding

Figure 2.38 Routing a stopped trench; using a strip to guide the router

3. Attach the guide, setting it to the correct width.

4. Make a trial cut on a piece of scrap.

5. Place the router on the timber to be cut, making sure the bit is over the edge and not touching the job. If a stopped housing joint is made using a plunge router, the bit should be over the starting point of the joint, with the router in the raised position.

6. Turn on the router and begin to move it slowly from left to right. If the bit begins to smoke or burn the timber, increase the rate of movement slightly.

7. Turn off the machine and wait for it to stop before removing the router from the job.

MAKING A DOVETAIL JOINT

It is possible to construct a dovetail joint using a router and a special dovetail jig. There are many jigs on the market, ranging from the simple to the elaborate. The latter jigs are multi-adjustable for different-sized joints. The one shown in Figure 2.39 will handle 300 mm wide and 8 to 25 mm thick timber. This type comes set up for a lapped dovetail joint for a drawer front, where the sides of the drawer are 12.5 mm. The Gifkins jig is designed for use with a bench mounted router as shown in Figure 2.39.

Note: It is a good idea to practice on a piece of scrap to check that the settings are correct.

Constructing a common dovetail joint using the Roger Gifkins jig:

1. Mark the inside face of all boards (i.e. inside faces of the box or drawer). It is also useful to mark the bottom edges.

2. With the dovetail cutter in the router, adjust the height of the router so that the top of the cutter is above the blue template by the thickness of the timber you wish to dovetail. Before turning the router on, check that the bearing on the cutter is at least 2mm above the table.

Figure 2.39 Dovetail jig simplifies construction of dovetail joints

Figure 2.40 Dovetail router bit is guided into jig by ball race.

3. Stand the board for the dovetail slots on its end above the straight fingers (as shown in Fig. 2.39), with inside face against the jig. Adjust the board sideways to get the layout of the joint that you want.

4. Clamp work securely in place. Make sure clamps are safely above the cutter.

5. Set the stop at the side of the workpiece and lock in place. Do not move the stop until both halves of the joint are cut.

6. With the template on the jig running against the bearing on the cutter, cut the dovetail slots.

7. Remove the board with the dovetails. Stand the board for the pins on its end above the tapered fingers (on the other side of the jig), with the inside face against the jig and its edge against the stop. Clamp securely in place. Once again, check that the clamp is safely above the cutter.

8. Fit the straight cutter to the router (adjust the height as in step 2 if necessary).

9. With the template on the jig running against the bearing on the cutter, cut the pins.

SAFETY RULES FOR THE PORTABLE ELECTRIC ROUTER

1. **General safety.** Observe the general safety rules given at the beginning of this chapter.

2. **Hearing protection.** Wear earmuffs or earplugs to guard against hearing damage that could be caused by the sound that comes from the high-speed motors used in routers.

3. **Dust.** Some form of dust extraction should be used, especially when routing custom or particle board as the adhesives used in their manufacture can be dangerous if inhaled over a long period of time.

4. **Adjustments.** Always unplug the power lead before fitting cutters or making any adjustments.

5. **Starting.** Make sure the workpiece is secure before switching on and hold the router firmly to overcome the torque or twisting action of the high-speed motor.

6. Lead. *Keep the lead free of the cutting area when the router is in use. It is often good practice to place the lead over the shoulder.*

7. Direction of cut. *Try to use the router so that it is working against the direction of the rotation of the cutter. That is, work from left to right on the edge facing the operator.*

PORTABLE PLANERS

The portable power planer is virtually a portable electric jointer. An electric motor is mounted in a lightweight metal or plastic frame and a barrel, fitted with cutters, is belt driven to produce a cutting action. Some planers have one spiral cutter blade but most have two straight blades.

Planers range in size from a small block plane weighing less than 2 kg with a 40 mm blade to the largest planer weighing 9 kg with a 170 mm cutter. The cutter head rotates at speeds ranging from 12 000 to 25 000 rpm.

OPERATING ADJUSTMENTS

The body of the planer is an integral part of the rear fixed sole and the cutter blades are set at this level. The front sole is adjustable and usually has marked calibrations or settings to indicate the depth of cut.

Figure 2.42. Using a larger power planer to plane a rebate. Note depth stop and fence to control size of rebate.

Figure 2.41 Small power planer.

Figure 2.43 Battery powered planer. Note use of fence to control ninety degree cut.

Some planers are supplied with guides or fences which allow the operator to plane rebates. An adjustable fence can also be attached to facilitate the planing of bevels.

Adopt a similar stance to that used for normal hand-planing, with feet comfortably apart and the body balanced. But here the similarity ends, as a firm and steady forward movement is required. Apply pressure to the front of the plane for the start of the cut and transfer it to the rear sole at the end of the stroke so that the plane remains level. Always allow the motor to reach full speed before proceeding with the cut.

SHARPENING AND INSTALLING PLANER BLADES

Most manufacturers supply a sharpening guide with their planers which supports the blade at the correct angle for honing. Some manufacturers also provide a grinding jig to grind the blades uniformly when they become gapped or excessively worn. The grinding and sharpening operations are basically the same as for handplanes. The plastic installation guide provided allows the spring-loaded blades to be installed level with the rear fixed sole of the planer.

SAFETY RULES FOR THE PORTABLE PLANER

1. **General safety.** *Observe the general safety rules given at the beginning of this chapter.*

2. **Cutting depth.** *Always check for correct cutter depth adjustment before making a cut.*

3. **Bevel.** *Check for the correct bevel adjustment before making a cut.*

4. **Blade replacement.** *Do not attempt to remove or sharpen the blade or cutters without carefully reading the manufacturer's manual.*

5. **Hearing protection.** *Always wear earmuffs or earplugs as the high-pitched noise from the machine could damage hearing.*

6. **Blade protection.** *Check timber for loose knots, or the presence of nails or foreign objects that could damage the blades.*

7. **Grain direction.** *Check grain direction to ensure a smooth cut.*

8. **Lead.** *Always keep the power lead out of the path of action. It is good practice to keep the lead over the shoulder to avoid contact with the machine.*

PORTABLE BISCUITING MACHINES

The biscuiting machine, or plate joiner, is a newcomer to the field of cabinet making and it has revolutionised the speed at which a job can be set out and assembled. It bypasses much of the layout work and many of the cutting operations associated with constructing traditional joints, such as dovetails, dowels and rebates.

It is basically a plunge-cutting circular saw with a guide for referencing the blade to the stock. Built-in stops are provided to allow the cut to be set in the correct position and at the correct depth for the biscuits (Fig. 2.44).

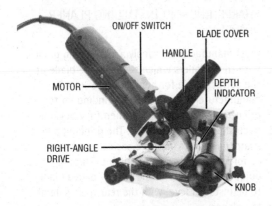

ON/OFF SWITCH
BLADE COVER
HANDLE
MOTOR
DEPTH INDICATOR
RIGHT-ANGLE DRIVE
KNOB

Figure 2.44 A biscuiting machine is basically a plunge-cutting circular saw

A biscuit or plate joint is a type of spline or 'floating tenon' joint. It is fast, accurate and safe to produce as well as being suitable for all types of wood and wood-based materials, such as plywood and particle board (Fig. 2.45). Slots are cut in both pieces to be joined and precut biscuits, or plates of compressed wood, are glued into the slots. As with dowel joints, the members or parts of a job can be cut to final size before joining, but the slots for biscuits are easier to fit and more forgiving than dowel holes. There are three different sizes of biscuits available for different widths and thicknesses of material.

The biscuits are made by pressing with a die so they are in a compressed form and should fit snugly into the slot (provided they are kept dry). When the glue is applied and the joint is cramped, the biscuits expand in the slots making

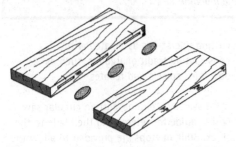

Figure 2.45 Biscuit or 'floating tenon' joint: an alternative to dowelling for edge joints

a strong joint. The slots locate accurately parallel to the face of the biscuits but are 1 or 2 mm longer than the biscuits, allowing the joint to be shifted along its length (Fig. 2.45).

Figure 2.46. Job clamped to the backing piece which serves to control position of slot; note slot and biscuit in foreground.

MAKING A BISCUIT JOINT

1. *Cut the timber or each member to size and place the two mating pieces together.*

2. *With a pencil, mark the centre of each face of the pieces where the biscuits are to be placed (usually about 50 mm apart).*

3. *Set the fence to locate the position of the slot, generally in the middle of the edge, and the depth of cut to correspond with the size of the biscuit used.*

4. *Separate the two pieces and the joint is ready to be cut.*

5. *Align the biscuiting machine with the marks on the job.*

6. *Switch the machine on and push it into the job, cutting the slot.*

7. *Repeat the operation until slots are cut at each pencil line.*

8. *Place adhesive in the slot and on the edge to be jointed and place a biscuit in each slot in one side of each mating pair.*

USES AND LIMITATIONS

The primary use of the biscuiting machine is for carcass construction. It can be used for making panels and frames or for cutting grooves. The biscuits can be used to edge-join boards, for example in table tops, or a leg or stile and rail. Mitre and butt joints can also be reinforced with biscuits. Shelving or partitions can be joined by cutting slots in the face of one panel and in the edge of the other.

Larger cross-sections of members to be joined may need the use of two rows of biscuits side by side.

The main disadvantages of the biscuiting machine are its inability to join very small or very large parts. Another problem to be considered is that the biscuit will not align accurately along the joint due to the slot allowing a small amount of adjustment.

The biscuiting machine can also be used to cut grooves or trenches.

SAFETY RULES FOR THE PORTABLE BISCUITING MACHINE

1. **General safety.** *Observe the general safety rules given at the beginning of this chapter.*

2. **Adjustments.** *Always disconnect the lead from the power point when making adjustments.*

3. **Securing workpiece.** *Make sure the workpiece is safely secured or held. Never try to cut very small pieces that cannot be securely held.*

4. **Hearing protection.** *Wear earmuffs or earplugs if the machine is to be used over an extended period of time.*

5. **Dust.** *Wear a dust mask if an efficient form of dust extraction is not available.*

QUESTIONS

1. *Describe the procedure to be followed when replacing a belt on the belt sander.*

2. *What advantage does the orbiting disc sander have over the normal rotating disc sander?*

3. *List the main types of portable electric drills available on the market and explain the principle of double insulation.*

4. *Explain the following operations when using the jigsaw: freehand cutting, ripping, crosscutting, plunge cutting.*

5. *List the principal safety precautions to be taken when using the portable circular saw.*

6. *What is the purpose of the plunge router? Explain how to preset the depth of cutter.*

7. *Some of the latest routers have electronic speed control. Explain the purpose of this option and list its advantages.*

8. *Why are some router bits made from high-speed steel and others from tungsten carbide?*

9. *The importance of correctly set and properly sharpened blades on a portable power planer cannot be over-emphasised. Describe the procedure for removing, sharpening and replacing planer blades.*

10. *The biscuiting machine is particularly suited for making floating tenon joints for edge jointing. Describe the steps you would follow to construct this joint when joining boards for a table-top.*

11. *Battery powered portable tools have definite advantages over 240 volt powered tools. List the advantages and limitations.*

SURFING THE NET

Woodworking Association

http://www.vicnet.net.au/~woodlink/tools.html

Chapter 3

Architectural Design and Building Construction

Architectural design is a complex process. Basically it is the designing of a building to meet the particular requirements of the prospective owner and covers the functional, structural and aesthetic planning.

FACTORS IN ARCHITECTURAL DESIGN AND PLANNING

Many small buildings and houses are designed by their owners or by builders, but most designs are prepared by architects. For large building projects architects work in close collaboration with structural engineers and other consultants such as lighting, heating, ventilation or air-conditioning experts. Builders work in close collaboration with the architect and his or her associate designers.

The design is shown on drawings collectively referred to as 'plans'. The first step is for the architect and owner to view the site and discuss their ideas about the building and the best use of the site. Sketch plans are then prepared and later, the finished plans. During these operations, problems relating to levels and contours, views, adjacent or possible new buildings, positions of rooms, doors and windows, roof shapes, and decoration are worked out. Some developments need to have an environmental impact study carried out to check the impact on the local environment.

In addition to the plans, written explanations about the materials to be used and work to be done are necessary. These are called 'specifications'. For large buildings a list of the quantities of materials required, prepared by quantity surveyors, is also provided to facilitate cost estimating and ensure that all those submitting estimates are working from the same information.

PLANS

As well as the arrangement of the rooms or parts of the buildings, all plans must show land contours, storm water drains and septic tanks if any, foundations, positions and sizes of doors and windows, roof shapes, fireplaces, chimneys and decorative features. No detail must be omitted.

House plans consist of a number of views—ground plan, front and rear elevations, at least one side elevation, a section made in the most convenient place to show foundations and structural members, and a block plan to show the position of the building on the site in relation to its boundaries and to any adjacent buildings (Fig 3.1). The position of the sun in relation to the building, an important consideration in designing to make the best use of natural warmth, light, and shade, is usually indicated by showing the north point on the block plan.

In the case of large buildings, plans for separate building operations such as concrete

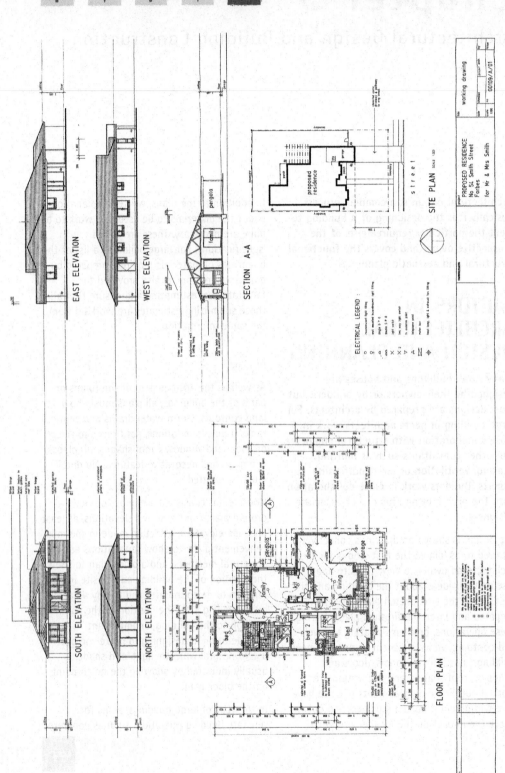

Figure 3.1 Example of plans to be submitted to local council

construction, plumbing, air-conditioning and electrical installation are required in addition to the main layout. Large-scale detailed drawings are sometimes necessary for special features of joinery or fittings, especially when these are to be made in separate workshops or factories.

Plans are usually prepared to a scale of 1:100 for main views, 1:50 for details, and 1:500 for the block plan (see, for example, Fig. 3.1). All necessary sizes are lettered on the drawings and, should there be any discrepancy, stated sizes are used in preference to measurements taken by means of a scale rule. Copies of the original plans are made by blueprint or photostat processes.

SPECIFICATIONS

Specifications amplify the drawings and include information such as proportions for concrete; size and kind of timber for construction, joinery and fittings; details of hinges, fastenings, locks, glazing, painting, electric light fittings and PC or 'prime cost' items such as a bath and stove for a home. They also set out standards of finish for the various trades. Specifications are arranged in order of work procedure from clearing the site and laying footings through to the finished building, with separate sections for each trade. They usually state the builders' responsibility to comply with all necessary by-laws and regulations and may also mention their obligations regarding public safety and clearing the site after construction is completed. A copy of the specifications is included with each set of plans.

APPROVAL OF PLANS AND SPECIFICATIONS

Before work can be commenced on a building, plans and specifications must be inspected and passed, and all copies stamped as having been approved, by the local municipal or shire council and the water, sewerage and drainage

authority. Other bodies must be consulted in some cases: in New South Wales, the State Planning Authority in cases involving proposed widening or alteration of roads, rezoning of land, and certain development projects; fire and police authorities in the case of public and commercial buildings, hotels and clubs; and electricity and gas supply authorities in the case of unusual or complicated installations, though generally inspection and test on completion is their only requirement.

Where large building or development projects are involved, preliminary drawings and details are submitted to the relevant authorities for 'permission in principle' before final plans are prepared. Permission in principle means that, other things being acceptable, the authorities have no objection to the idea. One example relates to a proposed shopping centre over a suburban railway station—permission in principle was given but the project did not proceed because the developers considered that the requirements of the local council were too restrictive and would add too much to the cost.

The local council retains one copy of the plans and specifications, the organisation providing finance may also require a copy and at least one copy is always kept on the building site and used for reference during construction.

Councils in New South Wales use the Australian Building Code as a basis for ensuring uniformity in building standards, and add other requirements to suit the particular needs of their areas. The code sets out required standards relating to construction, permissible floor areas and ceiling heights, natural light and ventilation, drainage, sanitation and hygiene, and position of buildings on the site.

The code requires that the minimum floor area of habitable rooms be 7.5 m^2. 'Habitable room' means a room normally associated with domestic living (other than a bathroom, laundry, toilet or the like). Dwellings and flats containing more than one habitable room are required to have a minimum floor area of 14 m² in one room and at least one other habitable room with a minimum of 11 m² (not counting any fully-enclosed kitchen). The position of toilet and kitchen in relation to other rooms and open area is also subject to regulation. Minimum height from floor to ceiling is 2400 mm, subject to certain standards of ventilation and insulation: Variations are also allowed in the case of rooms with a sloping roof, such as in 'A'-type and 'Cape Cod' buildings or attic rooms. To ensure proper light and ventilation, each room must be provided with an unenclosed window equal to at least one-tenth of the floor area. The large windows that are a feature of most modern homes are usually far in excess of this requirement. A window must not open on to a closed veranda.

RESPONSIBILITIES OF THE BUILDER

The builder usually contracts to construct the building for a certain stated price, or estimate. Any desired variation from the plans and specifications will affect this price, which must be adjusted at the final payment. As a considerable amount of money is involved, three or four payments are usually made as the work progresses, the builder receiving 70% to 75% of the value of the work actually completed at the time of each payment or 'draw'. Often 10% of the final payment is held for a period of 3 to 6 months as a guarantee that any faults, which may develop, will be corrected. Sometimes the contract is for the actual cost of labour and materials plus a percentage to cover the builder's supervision and organisation of the work. The contract may stipulate that the building must be completed within a stated time or a penalty paid. The builder as part of the contract pays all fees due to the various authorities.

Included among the builder's responsibilities are:

1. *To supply and have on hand all necessary materials of the standard required by the architect, including such items as roof trusses, joinery and fittings which may be pre-fabricated in the builder's or sub-contractors' workshops. Some sub-contractors, for example electricians, supply their own materials and most have their own tools and equipment.*

2. *To employ and organise the subcontractors so that all trades—concreters, plumbers, drainers, carpenters, electricians, plasterers, painters and so on—can carry out their parts of the work to required standards and without unnecessary delays. On large projects, a clerk of works is usually employed to act for the client or architect in checking standards of labour and materials, and it is the clerk's duty to be present at particular operations, such as pouring concrete.*

3. *To provide for proper vehicular access to the site.*

4. *To take any necessary safety precautions for the protection of the general public, such as the erection of hoardings.*

5. *To insure employees against accident and to comply with Department of Industrial Relations regulations.*

PHYSICAL FACTORS AFFECTING BUILDING DESIGN

Architectural design will be influenced by local government regulations concerned with the position of the building on the site, its distance from the boundaries and from adjacent buildings and the zoning of areas for certain types of building. Building covenants, imposed by local councils or land developers with the aim, like zoning, of controlling the overall quality of buildings in a given area, may place restrictions on material permissible for external walls and roof and on the minimum cost of building.

Natural conditions of sunlight, prevailing winds and general outlook must be considered so that they may be used to advantage. Existing buildings or the possibility of new ones and noisy or busy streets may present problems. Aspect is important and if you want a comfortable home it really does matter which way the house faces. The sun gives free heat (see Fig. 3.2) and cooling breezes do make a difference if the house is sited to take advantage of them. North and east are the most desirable aspects. This means that rooms on the north side will receive sunlight most of the day during winter. Some sites are more exposed to winds such as cold westerlies or southerlies. Design the house to shelter it from the worst of these winds: limit the amount of glass windows on that aspect or build the garage on that side.

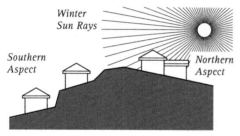

Figure 3.2 The sun provides free heat, and cooling breezes do make a difference if the house is sited to take advantage of them

Land contours or formations such as rocky outcrops often determine the design of a house. Although houses built on sloping sites can be made most attractive, very careful planning is necessary to make them easy to live in and to avoid high costs. Excavation or high piers are necessary if the house is to be on the one level, and use can be made of the space underneath for a garage, store or rumpus room An alternative is to build a split-level house (Fig. 3.3), which follows the land contours. The variation in levels is usually not more than a metre. Trees or rock formations are often used as a focal point, with the house designed around them (Fig. 3.4).

Figure 3.3 Split-level house

Figure 3.4 Tudor style home designed to blend into surrounding area

The design of footings is influenced by the nature of the land. Most foundations are of reinforced concrete laid as wide footings or a slab, and the width and thickness for particular loads are determined according to whether the soil is mainly sand, gravel or clay. Loose sand is unsatisfactory but dry, confined sand is good and so is gravel. Clay is affected by water and moves as it becomes wet or dry and for this reason is the least satisfactory. Solid rock makes an excellent foundation.

Particular care must be taken with footings on uneven ground and thickness or depth must be overlapped at all changes of level. Built-up soil must be allowed to compact or the footings should be taken down to solid ground by means of concrete piers or piles, the footings being carried across them in the form of beams. 'Floating' rock also needs special attention.

Faulty footings, although unlikely to cause a building to collapse, may cause walls to crack or floors to become uneven.

Wind and water can create building problems. Freestanding walls and tall structures or roofs must be designed to withstand severe wind pressures. Flat and skillion roofs must be tied down against the lifting effect of wind. Natural water seepage can cause dampness in walls and decay in timber. Provision must be made under wooden floors for cross ventilation by means of built-in ventilators.

Council regulations control storm water disposal—rainwater from roof, verandas or other areas cannot be discharged into sewer pipes. Where it is not possible to drain water to the street without affecting adjoining properties, rubble drains are necessary. These are trenches filled with broken bricks or stones and covered with earth to aid distribution and absorption of water.

Drainage work is costly and it is advisable to consider the position of kitchen, bathroom, toilet and laundry in relation to one another to avoid unnecessary expense. Where sewerage is available, all wastes must be connected to it. If no sewerage is available, waste water from kitchen, laundry, bath or shower may be run into sumps, but toilet waste must run into a septic tank. The run-off or effluent from septic tanks must be absorbed by the soil and must not run into drainage easements or natural watercourses. Clay is not a good absorbent and often requires special disposal trenches. Sand is good and stony ground usually satisfactory.

MECHANICAL FACTORS AFFECTING BUILDING DESIGN

Loads on structures are never constant as, in addition to structural and normal imposed loads, there are variable loadings due to natural causes such as wind, and extra loading as occurs when a building is crowded with people or extra goods. In some buildings the stress during construction is greater than the final imposed load and provision must be made for this, usually by some temporary means, which can be removed on completion of the work. In all structures there is continuous movement due to normal contraction and expansion of materials, which, in itself, can affect the structural strength.

The effect of loads is to create movement, such as twisting, bending or crushing, which causes tension, compression or shear stresses to operate within the structural members. Tension is a pulling stress, tending to pull the material apart; compression is a pushing or pressing stress and tends to press the material into a shorter length; shear is a sliding stress, where the parts tend to slide across each other (Fig. 3.5). Where possible, struts and braces are designed to resist compression rather than tension because joints in wood can be made more satisfactorily to resist compression.

Tension, compression and shear stresses, which act invisibly, are separate and independent but

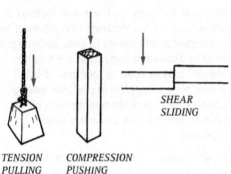
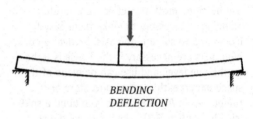

SHEAR
SLIDING

BENDING
DEFLECTION

TENSION
PULLING

COMPRESSION
PUSHING

Figure 3.5 Effects of stresses on structure

can operate within the one member at the same time, depending on its structural position. In a beam there is compression in the top half and tension in the bottom half every time it deflects and, if excessive, these forces can cause sliding or shear stress at their junction (Fig. 3.5). There is also a shearing force on a beam near its abutment (Fig. 3.6).

Certain terms used in structural design are explained as follows:

Stress is the force or load bearing per unit area on the structure measured in pascals (Pa). Stress operates within the structure itself in addition to being exerted by imposed loads. For this reason the load of the structure must be added to the superimposed load when the structure is being designed.

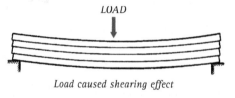

Load caused shearing effect

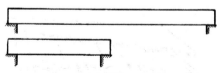

Beam twice as long will take half the weight

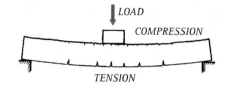

Effect of load on full beam supported at both ends

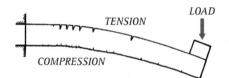

Effect of load on cantilever supported at one end

Figure 3.6 Effect of load on fully supported and on cantilevered beams

Loads are of two types, 'dead' and 'live'. The dead load is the actual weight of the structure and the live load the force imposed upon it, both usually expressed in newtons. These terms can be simply explained by examining a house. The dead load of the house is that which it possesses when it is completed but unfurnished and unoccupied. When the house is furnished and occupied, the added weight is the live load. Note that a live load is not necessarily one that moves. Moving loads bring further forces to bear due to transference of loads or pressure; also weight is a force which equals mass x gravity. Moving loads and vibration can cause structural breakdown more quickly than static loads (those that do not move) and the greater the speed at which the load moves, the greater the stress. An express train travelling over a bridge at 100 km/h would exert much greater stress than the same train at 30 km/h.

Strain is the amount of distortion in the material caused by the stress or load. It can also be brought about by changes in temperature without loading. Strain, being a ratio, is not expressed in units but as a number.

Deflection is the bending that occurs in a beam. A beam may deflect a certain amount due to its own weight. Further deflection takes place when load is applied. The amount of deflection and the load supported are directly proportional to the depth of the beam; this is why beams are placed on edge. Two beams of the same depth placed side by side will support twice the load of one beam and deflect half as much. One beam on edge, twice the depth of those mentioned, will support four times the load and deflect one-eighth as much (Figs 3.7, 3.8, 3.9).

While a beam may be strong enough to support a certain load safely, deflection could be excessive and cause, for example, cracking in a plastered ceiling or vibration in a floor. For this reason joists in a floor or ceiling are usually bigger in section than the load requires. Floor

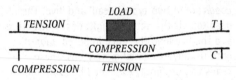

Deflection on fixed beam.

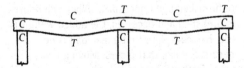

Deflection on continuous beam.

Figure 3.7 Deflection on fixed and continuous beams

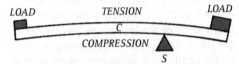

Lever on support becomes reversed beam.

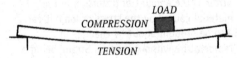

Load on beam causes deflection wherever it acts.

Figure 3.8 Illustrating two further principles in deflection

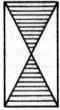

Section of rectangular beam, various intensities of bending

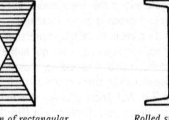

Rolled steel joist R.S.J. Economy of steel with greatest strength

This beam will support specified load and will deflect

Twice the thickness

Will support twice the load, deflect half as much

Twice the depth will support four times the load. Deflection one eighth

Figure 3.9 Relative load-bearing strengths

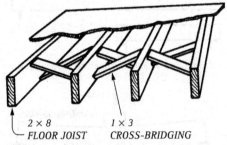

2 × 8
FLOOR JOIST

1 × 3
CROSS-BRIDGING

Figure 3.10 Herringbone strutting

joists are supported on plates and bearers, with brick piers to decrease the span and give the necessary stiffness. Ceiling joists cannot be supported in this way and therefore have a beam on top or, in the case of deep joists, are strutted between with rows of herringbone strutting (Fig. 3.10). The amount of deflection is stated as a ratio to the span and for a ceiling should not be more than 1:360.

Elasticity is the property of materials that enables them to return to their original shape after having been contracted, dilated or stretched. Some materials, rubber for example, are very elastic. Strips of rubber have this property to such a degree that they are called 'elastic bands'. All materials used in building,

even steel, concrete and glass, are elastic to a degree. Some timbers, such as spotted gum, are much more elastic than others, such as cypress pine.

Elastic limit is the limit of stress to which a material can be subjected and return to normal without displacing its particles. Beyond this point it will not return to its original shape and permanent strain and eventually breakage will result. Stretching a piece of elastic or bending a piece of thin wood can easily demonstrate this.

Modulus of elasticity is the ratio of stress to strain. Hooke's Law states that 'the strain is directly proportional to the applied stress'. With few exceptions elastic materials obey this law, which applies only before the elastic limit is reached. The letter 'E' represents the modulus of elasticity, also known as 'Young's Modulus' or the 'coefficient of elasticity'. There is a definite relationship between E and the strength of timber: the higher the value of E, the greater the strength.

Modulus of rupture is the load necessary to cause the material to break or rupture.

Factor of safety is the difference between a structure's actual load and the full load their members will support in theory. This factor is variable, depending on the type of structure and position of the member, but is usually between 1:5 and 1:10.

TIMBER AS A STRUCTURAL MATERIAL

Unlike metals, which can be melted and mixed to achieve predetermined qualities of strength and behaviour, wood is a naturally-produced material, which must be used with all its inherent advantages and disadvantages.

Some of the advantages of timber as a structural material are:

1. *good strength/weight ratio comparable to that of structural mild steel;*

2. *relative stiffness;*

3. *resilience—absorbs overloads without rapid strength deterioration;*

4. *low thermal conductivity;*

5. *low thermal expansion;*

6. *considerable natural beauty.*

Some of the disadvantages to be considered are:

1. *proneness to moisture absorption and drying out, even after seasoning;*

2. *tendency to change in sectional size and shape during seasoning;*

3. *tendency to split, warp, bend and bow;*

4. *susceptibility to attack by insects and fungi;*

5. *grain variation in the form of knots and twists and subsequent loss of strength;*

6. *in small sections, tendency to burn easily.*

Most of these problems can be eliminated or minimised by modern timber treatment methods.

Design techniques for steel and concrete cannot be applied to timber. Shrinkage is always a problem in joining structural members, and timber strength varies not only from species to species but also within species and even within one tree. Strength is affected by:

1. **Moisture content.** *Dry timber, accepted at 12% moisture content, is stronger than green timber.*

2. **Knots.** *These change grain direction and also become loose, decreasing the effective timber section.*

3. **Slope of grain.** *Maximum strength, obtained with the grain parallel to the length of the timber, decreases proportionately as the grain slopes away and becomes shorter.*

4. **Defects.** *These include gum veins or pockets, checks, shakes, splits, bow, twist, cup, blue stain and decay.*

5. **Rate of growth.**

To enable the design of timber structures using different species with similar mechanical properties, timber has been divided into four different strength groups, A, B, C and D, the strongest being A. These groupings are subject to the proper selection and quality control of timber used.

Research organisations are constantly testing timber strength and collating results to achieve some degree of standardisation and to enable timber to be supplied to strength specification. Timber is fed into an automatic 'stress-grading machine' which tests at intervals to determine correlation between modulus of rupture and modulus of elasticity. Grading done by computer is indicated by a coloured mark on the timber or by printing the actual stress value along the length of the piece.

Much strength is lost in timber used structurally if joints are not well made and of the correct type for the particular situation. Normal jointing methods and the usual type of fasteners, such as screws, nails, bolts and adhesives, have been proven satisfactory for normal building framing, but large structural work sometimes demands special techniques and equipment. Many types of connectors

have been devised for use in trusses and frames, all of which allow a certain amount of slip due to timber shrinkage (Figs 3.11, 3.12). Galvanised steel gang nails are used to strengthen the joints on a truss roof frame (Fig. 3.13).

Increasing use is being made of structural laminating, in which layers of timber, generally 20 mm thick, are glued together to form straight or curved beams and other structures (Figs 3.14, 3.15). A new development in laminated beams is pre-stressing, in which cables are fixed to chucks and strained through slots in the beams. Greater strength is attained without altering the modulus of elasticity.

Plywood, aesthetically pleasing and mechanically sound for structural components, has advantages over solid timber in being relatively free from warping, cracking and checking, and can be preformed to shape. It is also used for concrete forming or as a support during construction, as in the Sydney Opera

   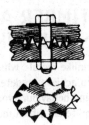

Figure 3.11 Types of timber connectors

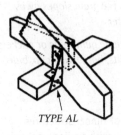
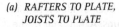

TYPE AL

(a) RAFTERS TO PLATE,
    JOISTS TO PLATE

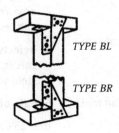

TYPE BL

TYPE BR

(b) STUDS TO PLATE

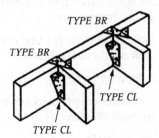

TYPE BR

TYPE BR

TYPE CL

TYPE CL

(c) JOISTS TO TRIMMER
    OR PLATE

Figure 3.12 Positions of framing anchors

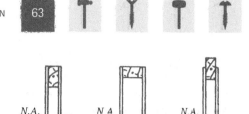

House. Bonding to an outer layer of aluminium or plastic makes it waterproof—Melbourne's Music Bowl has a cladding of 12 mm plywood with an outer skin of aluminium. Highly satisfactory beams of box pattern or I shape may be built up of plywood with solid timber runners. Box beams have been used for spans up to 30 m (Figs 3.16, 3.17).

Design the flange-web connection to have maximum allowable stress C—more efficient than B, while B is better than A.

Figure 3.16 Built-up box beams with plywood faces and solid timber runners

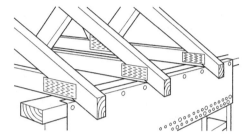

Figure 3.13 Gang nails on roof truss

Figure 3.17 An installation of plywood box beams

Figure 3.14 Laminated timber arches bolted to concrete bases

GENERAL BUILDING CONSTRUCTION

EXCAVATIONS

Building operations begin with the excavations, which are set out according to the building plan and may be for foundations only or for the overall building.

The site must first be cleared of trees, scrub and long grass and also of stumps and roots, which harbour termites. Proper levels must be taken with a dumpy level or straight edge and level (Figs 3.18, 3.19). It is usual to work from the high corner of the site and set out a peg known as the 'datum' peg for the base height, string lines being fixed to outline the foundation

Figure 3.15 Laminated beams supported on special concrete footings

levels (Fig. 3.20). It is most important that the building be set out square as well as level. Measuring diagonals or working with the rule of 3:4:5 can check this. Any multiple may be used—6:8:10, 12:16:20 and so on—but the longest possible measurements should be taken in order to obviate error.

Excavations must be level and stepped to allow for the fall of the ground. For brickwork, steps must be of a depth to suit courses (Fig. 3.21) and for accuracy it is usual to set out a rod marked with the courses, allowing for

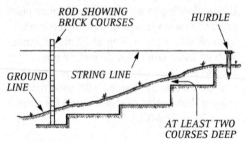

Figure 3.21 Excavation for brick or concrete footings

mortar joints. Excavations must be cut cleanly to width and in suitable ground may be used instead of timber forming as a frame for pouring concrete footings. Footings usually run through door openings.

FOOTINGS

Reinforced concrete is used as footings in most cases, the width and depth depending on the load, type of ground and the thickness of the walls. On soft or made-up ground, concrete piers or piles are used spanned by concrete beams (Fig. 3.22).

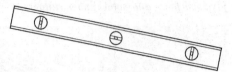

Figure 3.18 Spirit level from extruded aluminium

Figure 3.19 Small tube level hangs from chalk line

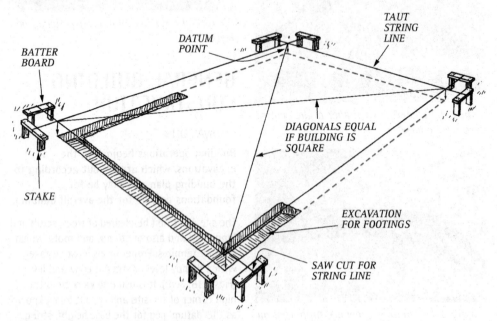

Figure 3.20 Setting out for foundations

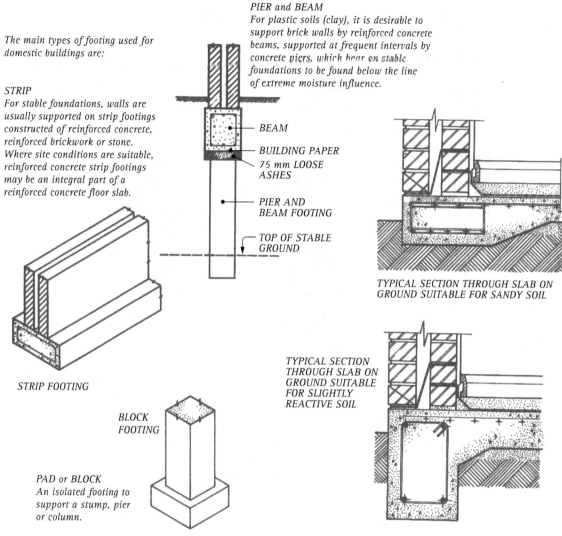

The main types of footing used for domestic buildings are:

STRIP
For stable foundations, walls are usually supported on strip footings constructed of reinforced concrete, reinforced brickwork or stone. Where site conditions are suitable, reinforced concrete strip footings may be an integral part of a reinforced concrete floor slab.

STRIP FOOTING

PIER and BEAM
For plastic soils (clay), it is desirable to support brick walls by reinforced concrete beams, supported at frequent intervals by concrete piers, which bear on stable foundations to be found below the line of extreme moisture influence.

BEAM

BUILDING PAPER
75 mm LOOSE ASHES

PIER AND BEAM FOOTING

TOP OF STABLE GROUND

TYPICAL SECTION THROUGH SLAB ON GROUND SUITABLE FOR SANDY SOIL

TYPICAL SECTION THROUGH SLAB ON GROUND SUITABLE FOR SLIGHTLY REACTIVE SOIL

BLOCK FOOTING

PAD or BLOCK
An isolated footing to support a stump, pier or column.

Figure 3.22 Types of footings

A concrete slab covering the complete building outline is sometimes laid, on which all walls and piers are built. This slab is usually bedded on level, smooth sand, with a polythene sheet spread over the full area under the concrete.

Brick in cement mortar is also used as footings, their size being determined by the load, type of soil and thickness of wall. Footings for piers may be of concrete or brick. Some states allow timber stumps as piers, but in New South Wales they are restricted to unimportant buildings such as sheds (Figs 3.23, 3.24). Galvanised iron caps, projecting at least 40 mm all round, are laid on top of all piers below plates and bearers. These act as a damp course and prevent access of termites to the timber.

To prevent water rising up the walls and causing decay in timber floor members, a

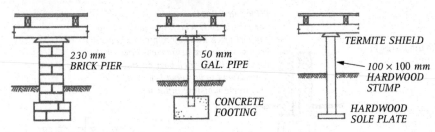

Figure 3.23 Piers under floor bearers

230 mm
BRICK PIER

50 mm
GAL. PIPE

CONCRETE
FOOTING

TERMITE SHIELD

100 × 100 mm
HARDWOOD
STUMP

HARDWOOD
SOLE PLATE

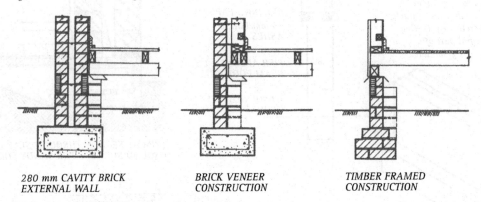

280 mm CAVITY BRICK
EXTERNAL WALL

BRICK VENEER
CONSTRUCTION

TIMBER FRAMED
CONSTRUCTION

Figure 3.24 Details of supports for bearers

damp-proof course of lead, aluminium or bituminous felt is laid below the floor line of all rooms. A similar course, laid so that the flashing comes right up to it, must be placed in chimneys and parapets above the roof line to prevent dampness soaking down.

FLOORS

In a house, the floors are made of concrete, a timber-based material such as particleboard or plywood, solid timber (generally tongue-and-grooved boards) or a combination of these materials.

Timber floors are built up with bearers (these generally being 100 × 75 mm on the edge) supported by the piers (the maximum spacing being 1800 mm centre to centre). Then the floor joists are placed at 90° and on top of the bearers (generally 100 × 50 mm on their edge spaced at 450 mm between centres) (Figs 3.23, 3.24). For brick construction, the floors are framed up between the walls of each room

and for timber, the floor is laid over the whole building with plates under all internal walls.

In a brick building, plates, bearers and joists are laid in position when the brickwork is at floor level and form a base for the scaffolding. In some cases the floorboards are cut in at this stage and temporarily laid face down until walls have been plastered, then turned over, cleaned and re-laid. In a timber or brick-veneer building, a platform construction is used where the flooring is laid over the whole area then the wall frames are fixed in place.

Floorboards for houses are usually made of cypress pine or tallow wood, but other timbers may be used. Boards to be finished in clear plastic must be carefully sanded, and in all cases nails must be punched. For ballroom floors or where a high-class finish is required, secret nailing is used. This requires boards that, to avoid open joints due to shrinkage, are not more than 75 mm wide, milled with a specially shaped tongue-and-groove. Each board is laid separately and nailed through the bottom of the tongue so that the nail holes do not show on the finished floor.

A special particleboard has been developed which uses a waterproof adhesive to join the wood particles and is therefore suitable for bathroom flooring. It comes in large sheets which cuts the laying time of the floor dramatically. Waterproof plywood is also being used in bathrooms, where it is coated with epoxy resin sealant after being laid and then tiled. Other types of materials used as bathroom flooring include suspended concrete and compressed fibre cement.

Timber is not often used for external floors as rain soaks in, especially on the ends, and alternating wet and dry conditions cause timber to deteriorate quickly.

Proper ventilation under floors (Fig. 3.24) is most important as rising dampness can cause decay and 'dry rot', a fine fungus which grows under damp conditions. A clearance of at least 200 mm under all floors to give access for inspection or repairs is required by ordinance, and if walls below the floor line enclose the building, an access door must be built into the wall.

Before concrete floors are laid, all necessary drainpipes must be installed. Concrete floors are sometimes used with a wooden floor overlaid on 50 × 50 mm supports nailed to the concrete with special nails. Floor coverings may be laid directly on concrete, composition floor coverings being cemented in place.

WALLS AND PARTITIONS

Walls and partitions may be of brick, concrete blocks, solid concrete, timber, or timber with an outer wall of brick, known as 'brick veneer'. Brick veneer (Fig. 3.25) offers the appearance and low painting requirements of a full brick house, but is less costly and has the advantage of easy fixing for internal walls.

Figure 3.25 Timber frame showing use of plywood bracing

Burnt bricks are made to a standard size of 230 × 110 × 76 mm or a modular brick size of 290 × 90 × 90 mm in a wide range of colours and several textures. Concrete bricks or blocks are made in a range of sizes, including standard brick size, and in several colours and face finishes. Larger sizes are made hollow to reduce weight and to allow concrete to be poured in if desired as a means of securing the blocks.

Brick walls are built in thicknesses that are multiples of 110 mm, the width of a single brick. Internal walls are 110 mm thick. External walls consist of two 110 mm walls with a 60 mm cavity between to prevent moisture soaking through. Galvanised-iron wall ties bind both walls together (Fig. 3.26). For very high walls or those carrying heavy loads the inner wall may be 230 mm or more thick. Concrete blocks can be made water-resistant and may be laid without a cavity. Damp-proof courses and flashing must be used as described in the section on footings (pp. 65–6).

Door and window openings are bridged with heavy angle iron, universal or I beams, concrete beams or brick arches. Whatever the material, the straight, horizontal bridging members over openings are called 'lintels'.

All walls must be laid in level courses with faces, angles and reveals kept plumb (Fig. 3.27). Bricks must be laid with mortar made from lime or cement mixed with sand and in a proper bond so that vertical joints are staggered. Single brick walls are laid in stretcher bond; that is, all bricks are laid end

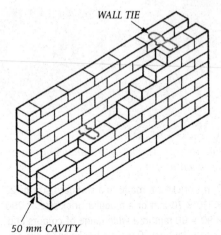

WALL TIE

50 mm CAVITY

Figure 3.26 Brick cavity wall in stretcher bond

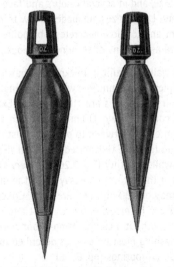

Figure 3.27 Metal plumb bobs are suspended on string to test vertical surfaces for plumb

to end with a half-brick at the beginning of each alternate course. Thicker walls of 230 mm or more may be laid in English, Flemish, 'garden wall' or other bonds (Fig. 3.28). Joints on the face of visible walls are finished by being struck, flushed, raked or grooved. Joints of walls to be plastered are not finished except for removal of excess mortar. A quick finish, called 'bagging', can be given to areas

such as the inside walls of a garage by spreading mortar with a trowel and rubbing it with a piece of bag or hessian or a stiff brush.

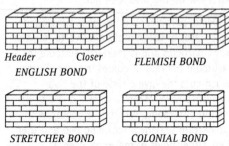

Header   Closer
ENGLISH BOND   FLEMISH BOND

STRETCHER BOND   COLONIAL BOND

Figure 3.28 Brickwork bonds, shown here in solid walls

Masonry walls may be plastered internally with cement or lime mortar, cement giving a much harder finish. If lime mortar is used, two coats are usual. The first, called 'floating', is finished with a wooden float to give a key for the second, called the 'setting coat'. This is a mixture of lime putty, plaster and sometimes clean sharp sand, and is finished to a smooth trowelled surface. Colouring may be added but the finish is generally left white to take paint, which offers a better range of colours.

The popularity of unplastered brickwork for internal walls is increasing, joints being finished as for external walls. Overburnt bricks, called 'clinkers', are also used both internally and externally to give a texture or to make a feature wall.

Timber walls are usually framed up on the floor joists, with top and bottom plates, studs, nogging and braces, and then raised into position (Fig. 3.25). For external walls all members are 100 × 50 mm, except braces, which may be 50 × 25 mm. Internal wall members are generally 100 × 50 mm but in some cases 75 × 50 mm is permissible. For external covering, or 'cladding', weatherboards and fibre cement are available in a number of sectional shapes, weatherboards being of hardwood, Baltic pine, Oregon, cypress pine and Monterey pine. Waterproof plywood and tempered hardboard, including a special 10 mm thickness, may also be used. Gloss paint or clear finishes may be applied.

Timber-framed walls are covered internally with plaster boards (gyprock), fibre cement, hardboard or similar materials. Wallpapers, which are applied with paste, may be used on any of these materials except patterned hardboard. They may be hung with matched patterns to cover the whole wall or in panels of feature patterns. Wallpapers have been developed which may be cleaned with a damp cloth, while some are washable with soap and water. Prepasted and precut papers are also available. Paint in a wide range of types and colours is a widely used finish for internal walls and woodwork.

In a timber home, bathroom walls to be tiled are usually lined with water-resistant fibreboard (villaboard) and tiles are glue fixed.

ROOFS

Roofs for houses are of three main types—pitched, flat, and skillion—the type selected depending mainly on the overall design of the house. Pitched roofs are more often used than the other two and are of gable or hipped type. Roofs of older houses are often very steeply pitched, mainly due to the influence of the Gothic style and of building practice in Europe, where a steep pitch is necessary because of snow (Figs 3.30, 3.31, 3.32, 3.33, 3.34).

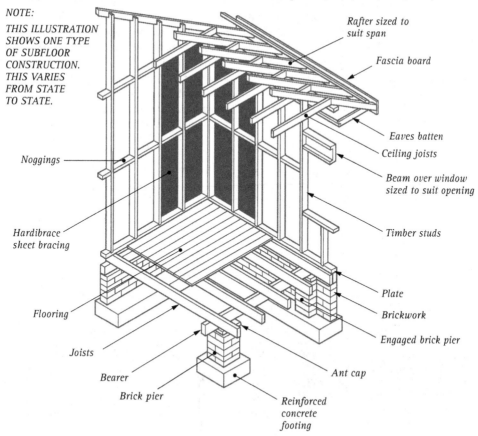

NOTE:
THIS ILLUSTRATION SHOWS ONE TYPE OF SUBFLOOR CONSTRUCTION. THIS VARIES FROM STATE TO STATE.

Rafter sized to suit span

Fascia board

Eaves batten

Ceiling joists

Beam over window sized to suit opening

Timber studs

Noggings

Hardibrace sheet bracing

Plate

Brickwork

Engaged brick pier

Flooring

Joists

Bearer

Brick pier

Ant cap

Reinforced concrete footing

Figure 3.29 Timber frame names and terms. One type of sub floor construction, this varies from state to state.

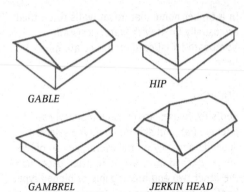

GABLE   HIP

GAMBREL   JERKIN HEAD

Figure 3.30 Types of roofs

Figure 3.34 Cape Cod style house with mansard roof

Figure 3.31 Cape Cod roof

Figure 3.32 Hip and gable roof

Figure 3.33 Steep gable roof with shed dormer. Note low-pitched roof on surrounding verandas.

Pitched roofs are framed by using the ceiling joists as part of each set of framing. Other members consist of rafters, ridge and hips and, to reinforce long spans of rafters, one or two purlins are used supported by struts from plates or ceiling joists on internal walls (Fig. 3.35). Struts must not be located on ceiling joists not supported by a wall, as roof movement would cause ceiling cracks. Collar ties are used where struts are impracticable and sometimes in addition to them (Fig. 3.36).

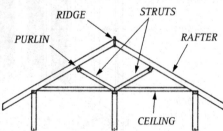

RIDGE   STRUTS

PURLIN   RAFTER

CEILING

PURLINS AND STRUTS ONTO INTERNAL WALL

Figure 3.35 Positions of purlins and struts

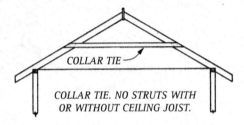

COLLAR TIE

COLLAR TIE. NO STRUTS WITH OR WITHOUT CEILING JOIST.

Figure 3.36 Collar-tie roof where absence of internal wall forbids installation of struts

These are fixed to the purlins, as collar ties do not give much support to the roof because of their angle to the load. Ceiling joists are fixed at 450 mm centres so that the feet of the rafters can be nailed to them to prevent spreading, and additional joists are necessary beside walls to support ceiling battens (Fig. 3.37). Cantilever joists square to main joists are sometimes used to receive rafters from the hips.

Oregon is commonly used for roof members as it is easy to work, light in weight and less likely to warp and twist than hardwood. Ceiling joists, rafters, struts and collar ties are 100 × 50 mm, purlins 100 × 75 mm, and hips and ridges 150 × 38 mm. Rafters are spaced 450 mm between centres for tiles, 600 mm for corrugated fibre cement and 760 mm for galvanised corrugated iron. To take these coverings, battens are nailed to rafters. Battens for tiles are 38 × 25 mm, spaced to suit the tiles (usually about 340 mm); for fibre cement, 75 × 30 mm, spaced not more than 760 mm between centres; and for galvanised iron, 75 × 25 mm, spaced not more than 900 mm between centres (Figs 3.35, 3.36).

Flat roofs are not made exactly level but must have a minimum fall of 1 in 40 achieved by sloping the joists or by fixing purlins of varying depths above level joists.

Skillion, or 'lean-to', roofs (Figs 3.39, 3.65) require a pitch or fall of not less than 1 in 12, depending on the type of covering material used. This pitch gives sufficient run off to prevent rain seeping through side or end laps.

Construction for both flat and skillion roofs over short spans are simple, requiring 100 × 50 mm rafters or joists. Boarding is used over flat roofs and 75 × 25 mm battens are used for skillions, the underside often forming the ceiling of the room. Over wide spans, and on a flat roof used as a sun area and thus carrying a superimposed as well as a structural load, deeper joists or rafters up to 300 × 50 mm are required, as deflection must not exceed 1:360. Stiffening in the form of herringbone strutting, 50 × 50 mm timber fixed diagonally each way in rows between joists (Fig. 3.10), is used if intermediate beams below joists or rafters are impracticable or are very widely spaced. However, intermediate cross-beams of timber or universal beam, or welded trusses of steel rods, built into sidewalls, are used where possible. Particularly suited to skillion roofs is the use of heavy beams or trusses with purlins and no battens. Story posts, which permit the use of smaller-section structural members, may support long spans.

Both flat and skillion roofs may have open or lined overhanging eaves, as for pitched roofs,

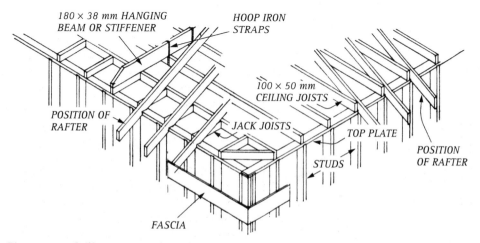

Figure 3.37 Ceiling construction

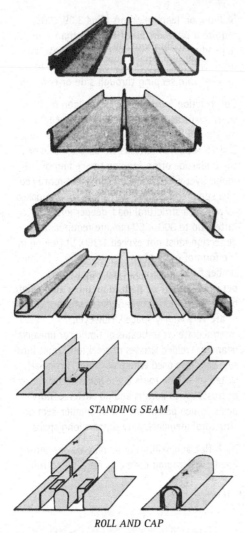

STANDING SEAM

ROLL AND CAP

Figure 3.38 Typical ribbed-metal roofing sheets and jointing methods for flat sheets

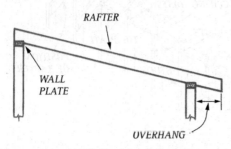

RAFTER

WALL PLATE

OVERHANG

Figure 3.39 Skillion or lean-to roof

and they may be framed between walls or parapets, in which case box gutters and wall flashing are necessary (Fig. 3.40).

Trussed roofs

In house construction, roofs are normally framed up without the use of trusses since internal walls can be used for strutting; but where a large span without internal supports has to be bridged as in a factory, store, or public building, a trussed roof is necessary. A truss, also called a 'principal', is a framework built up of materials big enough in section to support other roof members. Depending on span, it may be made of timber, steel or a combination of both, known as 'composite' (Fig. 3.41). The use of timber and composite trusses is limited to spans up to 14 m.

Trusses are usually placed 2.75 to 4.5 m apart and the rest of the roof is framed up on them. Purlins, common rafters and battens take the roof covering which is normally galvanised corrugated iron, clip-lock galvanised iron, clip-lock aluminium or corrugated fibre cement. In factories and similar buildings, ceilings are seldom lined so roof members are exposed.

The design of all roofs is governed by the tension and compression stresses bearing on the members. Rafters, beams and posts are in tension while braces and struts are in compression (Fig. 3.42).

Traditional types of timber trusses include king-post (Fig. 3.42) for spans up to 9 m and queen-post for spans up to 14 m. These are framed together with notched, tenoned and bridle joints secured with bolts, metal plates and straps.

The more modern nailed trusses include the Howe truss (Fig. 3.41), which is like a combination king-post and queen-post truss; the web-framed truss, really a beam built up as part of a sawtooth roof; and laminated timber built up into a truss, with plywood gussets and/or timber connectors for joining members. (See Figs. 3.11 to 3.15.) These trusses are spaced 1 m apart with common rafters between.

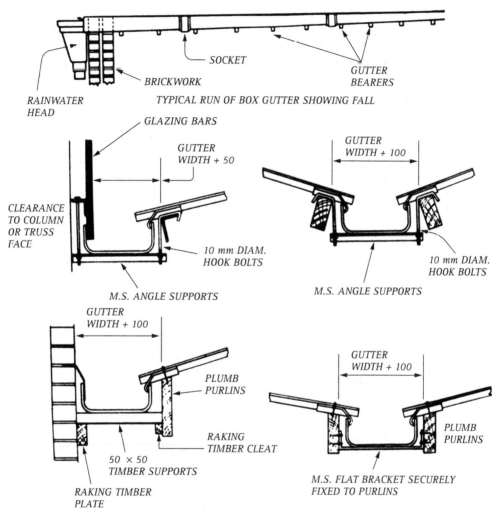

SOCKET

BRICKWORK

GUTTER
BEARERS

RAINWATER
HEAD

TYPICAL RUN OF BOX GUTTER SHOWING FALL

GLAZING BARS

GUTTER
WIDTH + 50

GUTTER
WIDTH + 100

CLEARANCE
TO COLUMN
OR TRUSS
FACE

10 mm DIAM.
HOOK BOLTS

10 mm DIAM.
HOOK BOLTS

M.S. ANGLE SUPPORTS

M.S. ANGLE SUPPORTS

GUTTER
WIDTH + 100

GUTTER
WIDTH + 100

PLUMB
PURLINS

PLUMB
PURLINS

RAKING
TIMBER CLEAT

50 × 50
TIMBER SUPPORTS

M.S. FLAT BRACKET SECURELY
FIXED TO PURLINS

RAKING TIMBER
PLATE

Figure 3.40 Typical methods of supporting box gutters

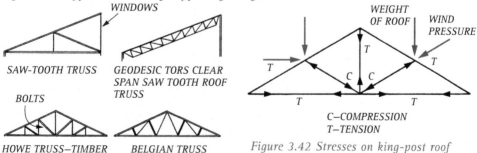

WINDOWS

SAW-TOOTH TRUSS

GEODESIC TORS CLEAR
SPAN SAW TOOTH ROOF
TRUSS

BOLTS

HOWE TRUSS–TIMBER
OR COMPOSITE

BELGIAN TRUSS

Figure 3.41 Typical steel and composite
roof trusses

WEIGHT
OF ROOF

WIND
PRESSURE

T

T

T

C C

T

T

C–COMPRESSION
T–TENSION

Figure 3.42 Stresses on king-post roof

Timber has the advantage of easy fixing for the roof covering, so composite trusses incorporate timber members with steel tension members, the latter usually in the form of long bolts (Fig. 3.14).

Steel is most commonly used for large trusses, though there is an increasing use of laminated timber for this purpose. Angle iron, which is easily joined, is used in the main section of steel trusses, and joints may be bolted or welded. Universal or I beams form the tie beams over wide spans, and lattice beams are often employed, particularly in sawtooth roofs.

Sawtooth roofs are often used on factories as the front end of the 'tooth' can be glazed to admit more natural light, an effect that is further increased when the end is at right angles to the roof pitch. Two types of trusses are used, the main truss spanning the building and the triangular-shaped trusses which it supports and which form the pitch of the roof. The disposal of rainwater from such a large roof area presents a problem: box or tapered gutters are required and, because these run across the building, any fault in construction or flashing or any blockage can cause leakage to the inside.

Mansard roofs (Fig. 3.34) have two slopes on each side and are used when more space is required above ceiling level.

Roof plumbing

Before roof coverings are fixed, guttering must be fixed to eaves and to valleys between the slopes of the roof. Flat roofs sometimes require box gutters (Fig. 3.40).

Gutters are commonly of galvanised iron in quadrant pattern from 75 to 125 mm in width, supported by galvanised-iron brackets. Copper, aluminium and fibre cement gutters are also available and one type of guttering is specially shaped to form a fascia as well. All gutters must be fixed with proper falls to round, square or rectangular downpipes of the same material. Joints are soldered or cemented with plastic according to the material.

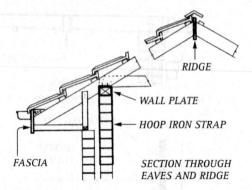

Figure 3.43 Roof anchorage and eaves treatment

Roof covering materials

Tiles in terracotta or concrete are used on pitched roofs. Terracotta tiles have a series of grooves and checks to prevent leaks, and if laid with a pitch of at least 1:4 to give good run-off they make a very satisfactory roof. Concrete tiles are not kiln-burnt during manufacture as are terracotta (clay) tiles and hence retain their shape without warping. Since this ensures a closer fit, their groove and check system is simpler. Hips and ridges are covered with special ridge-capping tiles bedded in lime mortar and pointed up with cement mortar coloured to match the tiles. Valleys are boarded and fitted with special galvanised-iron gutters and tiles are cut to afford a straight valley line.

Corrugated fibre cement (Fig. 3.44) is made in corrugation sizes of 140 and 75 mm and lengths from 1200 to 3000 mm. It is fastened to battens on pitched or skillion roofs by galvanised roofing screws with a galvanised-iron and rubber washer or lead washer under the head to prevent leaks. Very little pitch is necessary, though sealing with special mastic may be advisable at ridge and similar joints. Because of the brittle nature of the material, wire mesh is required under the sheets on a very large roof to make it safe to walk on.

Galvanised corrugated iron (Fig. 3.44), in 24- and 26-gauge and lengths up to 3000 mm, is

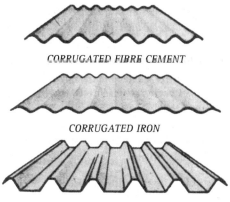

CORRUGATED FIBRE CEMENT

CORRUGATED IRON

ALUMINIUM ROOF DECK

Figure 3.44 Corrugated and ribbed roof
coverings

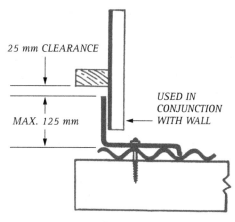

25 mm CLEARANCE

MAX. 125 mm

USED IN
CONJUNCTION
WITH WALL

Figure 3.46 Roof flashing for timber
construction

fixed to battens by roofing screws with a lead
washer under the head. Side lap is 1½ or 2
corrugations and minimum end lap is 200 mm.
Ridge capping covers ridges and hips, and
overhang at barges is finished with specially
made rolls. Though this material requires very
little fall, laps hold moisture and may be
painted to prevent rust.

Ribbed galvanised iron and aluminium (Fig. 3.38),
suitable for flat as well as pitched and skillion
roofs, come in long lengths which minimise
joints. Fixing methods vary, but roofing screws
and special clips are the most common. It is
important to use galvanised screws and fittings
with galvanised iron, and aluminium with
aluminium to prevent 'galvanic action'.

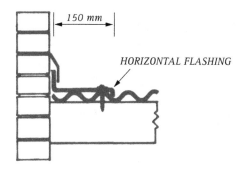

150 mm

HORIZONTAL FLASHING

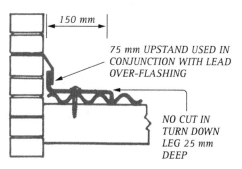

150 mm

75 mm UPSTAND USED IN
CONJUNCTION WITH LEAD
OVER-FLASHING

NO CUT IN
TURN DOWN
LEG 25 mm
DEEP

Figure 3.47 Roof flashing for brickwork

Figure 3.45 Concrete tiled roof and
brickwork blend into surrounding bushland

Fibreglass and plastic materials (with narrow
or wide corrugations) are applied either as
complete roof covering on patios and
verandas or over certain areas of the roof to
admit light. Laps and methods of fixing are as
for other corrugated materials except that a

plastic washer is placed beneath a metal washer under the screw head.

Roofing terms

The following terms are commonly associated with roof construction and are given with their meanings.

Eave—*The overhang of the roof beyond the wall. May be open but is now usually enclosed (Fig. 3.48).*

Soffit—*The underside of the eave or gable overhang, usually lined.*

Fascia—*The board covering the ends of the rafters, to which the guttering is fixed.*

Tilting piece or batten—*in a tiled roof, the extra-thick batten at the eave which raises the bottom of the first tile.*

Plumb cut—*The angle at which the rafter is cut where it joins the ridge or hips, at wall plate and end of overhang.*

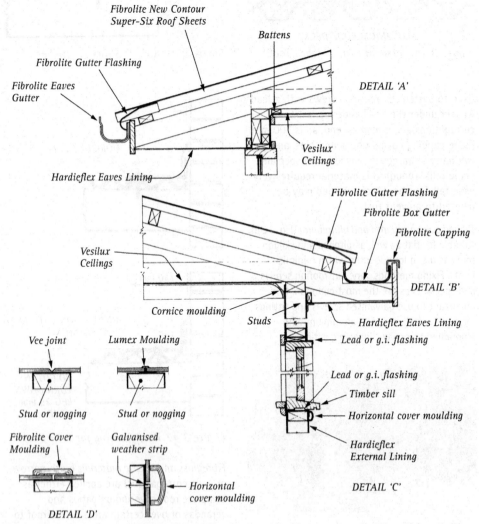

Figure 3.48 Construction details for timber frame with fibre cement cladding: (a) and (b) roofing and guttering; (c) treatment at window head and sill; (d) jointing of wall sheets

Edge cut—*The angle at which the rafter is cut across its edge; 90° for rafters joining the ridge, slightly less than 45° for those joining the hip.*

Birdsmouth—*The cut made on the rafter to fit over the wall plate or pole plate.*

Wall plate—*The plate, usually 100 × 75 mm, on top of the last course of bricks to which ceiling joists and rafters are nailed.*

Pole plate—*The plate on top of the ceiling joists to which rafters are fixed.*

Common rafter—*The rafter joining the ridge.*

Jack or creeper rafters—*The rafters joining the hip or valley.*

Hip cut—*The cut made on the rafter where it joins the hip or valley. Plumb cut is the same as for common rafters but edge cut is less than 45°.*

Barge—*The board fixed to the outer side of the gable overhang.*

Verge—*The tiles or other covering above the barge.*

CEILING LININGS

Ceilings may be lined with any of the materials mentioned for internal timber-framed walls, but fibrous plaster either plain or patterned, with a cornice, is the usual covering.

INSULATION

It has become the practice to insulate walls of timber homes and ceilings of all types in order to keep the building cool in summer and warm in winter. Insulation consisting of a bitumen-impregnated paper with thin aluminium laminated to each side is nailed to the outside of wall studs before the outer wall covering is fixed, and is installed beneath the tiles of a roof. In roofs it is called 'sarking' and, in addition to providing insulation, prevents rain blowing through on to the ceiling. Fibre insulation, either loose or made into blocks known as 'bats', can be placed in

walls but is more generally used as ceiling insulation. A layer at least 50 mm thick is laid on the ceiling between the joists, loose fibres being spread by hand or blower.

CONCRETE AS A BUILDING MATERIAL

Concrete has become one of the most important building materials. It is a form of artificial stone made by cementing together, with a paste mixture of cement, sand, and water, pieces of hard material called aggregate. Each piece is surrounded by the cement paste with no voids, or spaces.

Aggregate usually consists of gravel, crushed stone or blast-furnace slag. Ashes, broken brick or broken sandstone are also used but are less satisfactory. Furnace residue is liable to set up a chemical action with steel reinforcement and is not recommended for reinforced concrete.

TYPES OF CONCRETE

Reinforced concrete is used in bulk form for such work as machinery beds and heavy construction. Of high compression strength, concrete possesses low tensile strength which, for structures such as walls, floors, beams, roofs, bridges and silos, must be increased by the addition of reinforcement. This is in the form of mild steel rods, usually deformed to increase the grip of the concrete and made up into a mesh or grid for floors (Figs 3.49, 3.50).

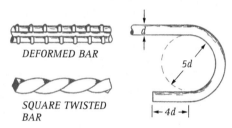

DEFORMED BAR

SQUARE TWISTED BAR

Figure 3.49 Types of steel reinforcement for concrete

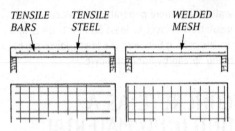

TENSILE BARS   TENSILE STEEL   WELDED MESH

Figure 3.50 Reinforcement for concrete slab floors

Reinforcement must be placed as near as practicable to the bottom of the concrete, that is, in the area of greatest tension. To resist the shearing stress near the abutments of beams, the ends of the reinforcement are bent up at about 45°.

Prestressed concrete is of two types, pretensioned and post-tensioned. For pretensioning, reinforcement in the form of high-tensile steel wire is set up and tensioned and the concrete is then cast around it. The tension is released after the concrete has set, compressing it and increasing the tensile strength. This method is particularly suitable for factory production of concrete components. Post-tensioning is more suitable for large structural members. Beams are cast with conduits along their length. When the concrete has cured, high-tensile steel wire anchored at the ends with special cones is stretched in the conduits using special jacks. The conduits may be filled with thin grout.

Lightweight concrete contains special lightweight aggregates such as expanded shale, foamed slag and natural pumice. This results in a great reduction of total structural weight, which allows a reduction in the size of supporting members. This in turn achieves a further reduction in total weight, an increase in useable space and a considerable lowering of cost. Lightweight concrete was used in most of the construction of the Australia Square building in Sydney.

STRENGTH

The strength of concrete depends on the strength of the aggregate, the proportions of sand and cement, the quantity of water, the thoroughness of mixing, and proper curing.

Proportions

For large structural work the required strength of the concrete is carefully worked out by structural engineers and, for accuracy, the proportions of the mix are measured by weight. For smaller projects, it is usual to measure proportions by volume, the aggregate forming the greater part. In determining the quantity, it must be remembered that the sand and cement paste fills the voids between the pieces of aggregate and adds little to the total volume. Buckets are generally used to measure quantities for small projects. The following are typical proportions for various types of work:

1. *Footings for brick buildings, beams, steps, retaining walls, suspended floor slabs—4 parts aggregate, 2 parts sand, 1 part cement.*

2. *Footings for timber buildings, slabs on the ground or filling, paths—6 parts aggregate, 3 parts sand, 1 part cement.*

3. *Paving subject to much wear—3 parts aggregate, 2 parts sand, 1 part cement.*

The thickness of concrete is important, 300 mm being usual for footings and 100 mm for supported slabs, such as small garage floors.

Quantity of water

Water should be just sufficient to ensure a workable mix and to hydrate the cement. The quantity of water is usually stated per bag of cement, 18 litres per 40 kg bag being a low ratio and 32 litres per bag a high ratio. The more water used, the weaker the compression strength of the concrete. Excess water tends to form in pockets and evaporate, leaving voids or spaces. It may also leak out causing a clear track to form—buildings have been known to fail structurally because of this.

Mixing

Whether concrete is mixed by hand or machine, thorough mixing of all components is most important. Mixing by hand is satisfactory for quantities up to a cubic metre, and should be done on a clean watertight surface. The usual method is to spread the aggregate on the mixing area and add the cement and sand, which may be premixed. The dry mixture should be turned over thoroughly until it appears even in mix and colour—this will take at least two complete turns. Water is then added, preferably with a fine spray, and the mixture turned over at least twice until a uniformly wet mixture is obtained. Machine mixing produces a more uniform concrete. It is difficult to keep the machine clean, but some aggregate and water rotated in the mixer first will clean the barrel.

The order of feeding the materials in may be: aggregate and water first, then sand, and cement last; or water first, then cement, and sand and aggregate last. In either case the mixing time is most important—at least two minutes' rotation time should be allowed after all components have been put in the barrel. A good mixing speed is 20 rpm. If rotation is too fast, centrifugal force throws the mixture to the outside of the barrel rather than mixing it.

Ready-mixed concrete is used for most structural work. It can be obtained mixed to specifications and is sold by the nearest metre to 0.2 m³.

Curing

Concrete sets touch-hard within hours but its ultimate strength is achieved after curing from 7 to 28 days, depending on the mix and weather and other conditions. Too-rapid drying out affects the strength of the concrete, so it should be covered with bags or plastic sheeting. In hot weather, spraying with water may be necessary. Plastic sheeting is often spread over the ground to prevent water absorption.

Formwork

Concrete must be placed in some kind of frame, called a 'form', to restrict it to the required size and shape. For large jobs, forming is a costly and time-consuming operation and special equipment is used to expedite the work. In structural work, such as floors above the ground, forming must be strong enough to take the weight of the workers as well as that of the wet concrete.

For small work, such as garage floors or paths, forms are usually planks held in position by wood or metal pegs driven into the ground (Fig. 3.51). Care must be taken to set up all forming level, plumb and to the correct size, and parts must be secured against distortion by the concrete during placing. Forms should be left in position long enough for the concrete to set and no superimposed weight put on it until curing has taken place. To prevent concrete sticking to forms, they may be painted with form oil.

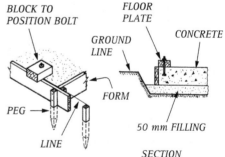

Figure 3.51 Details of corner of formwork for garage floor

Waterproof plywood and hardboard are used to impart a smooth finish to the concrete and for a special textured finish a board with a decorated surface, or even rough-sawn timber, is used for formwork.

PLACING

Concrete should be placed into position rather than poured, since pouring from a height greater than one metre tends to cause uneven density. Concrete should be compacted and

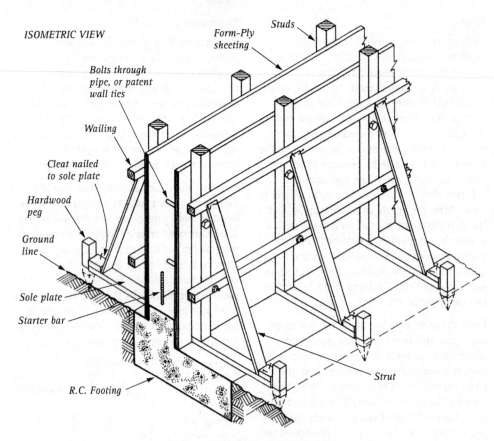

ISOMETRIC VIEW

Studs

Form-Ply sheeting

Bolts through pipe, or patent wall ties

Wailing

Cleat nailed to sole plate

Hardwood peg

Ground line

Sole plate

Starter bar

R.C. Footing

Strut

SMALL RETAINING WALL FORMWORK

Figure 3.52 Typical formwork for concrete wall

consolidated, particularly around reinforcement and edges, by poking with a steel rod or tamping with a wooden rammer. In large structural work, mechanical vibrators are used for this purpose.

Work should be screeded, or levelled, off at the top to an even height. For a floor, sloping across the entire surface with a long piece of wood will give an even finish but too much screeding will bring finings (fine pieces of stone) and water to the surface.

JOINTING

Long stretches of concrete, such as paths, tend to crack across the width owing to

contraction. To prevent this, joints are made about 2000 mm apart to break up the length, with tar-impregnated caneboard or tarred paper dividing the sections. Sometimes the path is laid by first filling alternate sections, then completing the remainder when the first concrete has set.

FINISHING

Concreting is finished with a wooden float or steel trowel. A wooden float (Fig. 3.54) gives a rather rough finish suitable for a garage floor or a path, where a smooth surface is liable to become slippery. For a smooth finish a steel trowel is used (Figs 3.55, 3.56).

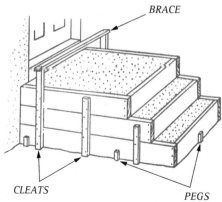

Figure 3.53 Formwork for concrete step construction

Figure 3.54 Wooden float for concrete

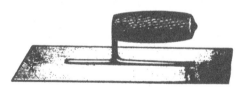

Figure 3.55 Steel finishing trowel

Figure 3.56 Another type of steel trowel for finishing concrete

When setting is just right, a time known in the trade as 'going off', the surface can be finished very smooth. Steel-trowelling has the effect of bringing water to the surface—dry cement sparingly sprinkled over the surface helps to dry it out. Conversely, if the surface is too dry, water splashed on with a brush makes it workable.

Edges and joints are given a neat finish with a metal jointing and edging tool.

TOOLS AND EQUIPMENT

The right tools and equipment are important. Most work calls for screeds, spirit levels, plumb bob, string line, large square, wooden or metal pegs, square-nosed shovel, wooden float, steel trowel, edging tools, hammer and nails, stiff brush, bucket and hose and, possibly, a rubber wheeled barrow.

Cement and concrete set quickly and are therefore difficult to remove from tools. All tools should be cleaned promptly by scraping off excess cement and then washing in water.

PRACTICAL CONSTRUCTION OF A SMALL BUILDING

Planning, the necessity to seek approval from the relevant building authorities, preparation of the site and the placing of footings has been discussed earlier in this chapter. This section describes the actual setting out and construction of a small building.

WALL FRAMING

(Study Figs 3.57 to 3.64.) Each wall is set out separately because of the differences in position of doors and windows. Set out the overall length of one plate, thickness of walls, positions of doors and windows, then intermediate studs. The wall covering must be considered, particularly if it is to be fibre cement or other sheet material, so that standard-width sheets can be used as far as possible to avoid cutting and unnecessary waste. If the building is to be lined, extra studs will be needed at the corners. Set these out, halving joints at corners and gauging depth of trenches, then cut joints and trenches.

Set out the length of one stud and use it to mark the others. Alternatively, nail two blocks

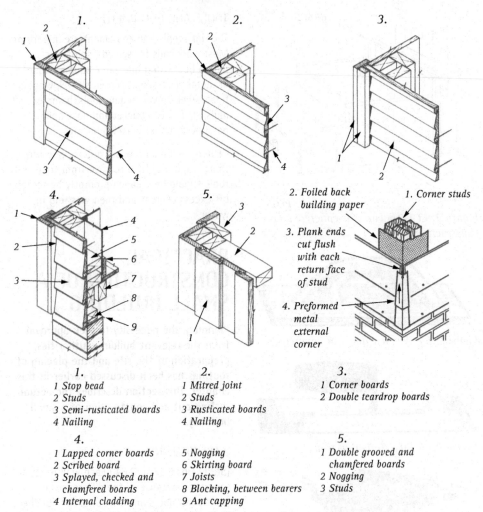

2. Foiled back
   building paper

3. Plank ends
   cut flush
   with each
   return face
   of stud

4. Preformed
   metal
   external
   corner

1. Corner studs

1.
1 Stop bead
2 Studs
3 Semi-rusticated boards
4 Nailing

2.
1 Mitred joint
2 Studs
3 Rusticated boards
4 Nailing

3.
1 Corner boards
2 Double teardrop boards

4.
1 Lapped corner boards
2 Scribed board
3 Splayed, checked and
  chamfered boards
4 Internal cladding

5 Nogging
6 Skirting board
7 Joists
8 Blocking, between bearers
9 Ant capping

5.
1 Double grooved and
  chamfered boards
2 Nogging
3 Studs

Figure 3.57 Details for finishing wall cladding at angles and openings

Figure 3.58 Special Harditex base sheets
for textured coatings

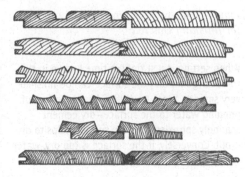

Figure 3.59 Typical sections of internal
lining boards

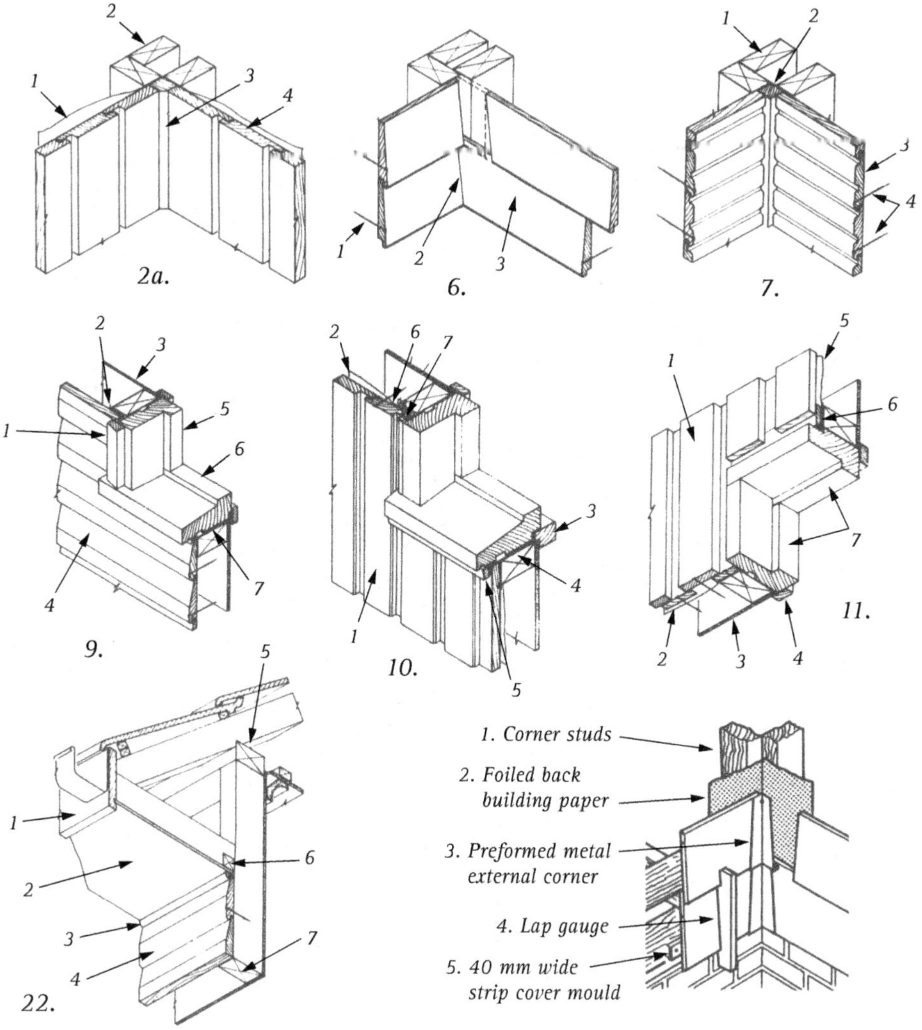

2a.	6.	7.	9.
Re-entrant corner:	1 Nailing	1 Studs	1 Storm mould
1 Sarking	2 Scribed joint	2 Stop bead	2 Flashing
2 Studs	3 Splayed, checked	3 Double Log	3 Internal wall lining
3 Corner mould	and chamfered	Cabin boards	4 Semi-rusticated boards
4 Vertical cladding	boards	4 Nailing	5 Window jamb
			6 Window sill
			7 Flashing

10.	11.	22.	
1 Double splayed 'U'	1 Square sawn boards	1 Fascia board	
joint ship lap boards	and battens	2 Asbestos cement coves	
2 Sarking	2 Sarking	soffit lining	
3 Window board	3 Internal wall lining	3 Quadrant	
4 Flashing	4 Architrave	4 Semi-rusticated boarding	
5 Timber mould	5 Storm mould	5 Wall plate	
6 Jamb flashing	6 Flashing	6 Fixing plate	
7 Storm mould	7 Window framing	7 Stud	

1. Corner studs

2. Foiled back building paper

3. Preformed metal external corner

4. Lap gauge

5. 40 mm wide strip cover mould

Figure 3.60 Corner and window joint details in timber construction

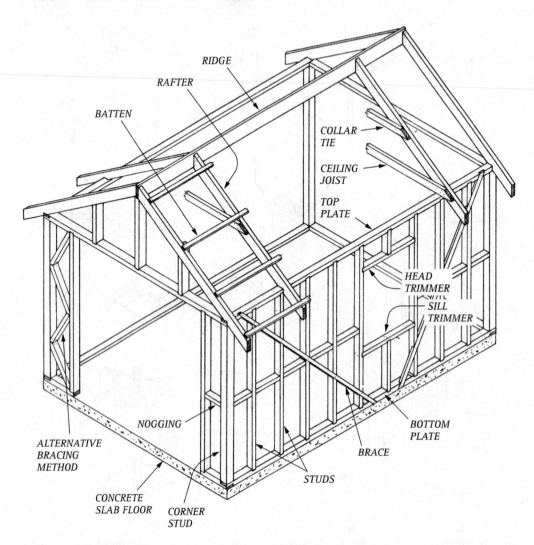

RIDGE

RAFTER

BATTEN

COLLAR
TIE

CEILING
JOIST

TOP
PLATE

HEAD
TRIMMER

SILL
TRIMMER

NOGGING

BOTTOM
PLATE

ALTERNATIVE
BRACING
METHOD

BRACE

STUDS

CONCRETE
SLAB FLOOR

CORNER
STUD

Figure 3.61 Timber framed garage on concrete foundation

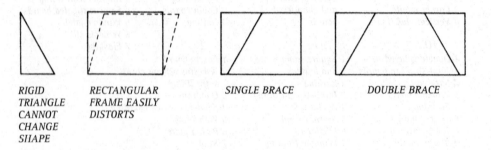

RIGID
TRIANGLE
CANNOT
CHANGE
SHAPE

RECTANGULAR
FRAME EASILY
DISTORTS

SINGLE BRACE

DOUBLE BRACE

Figure 3.62 Theory of bracing frames

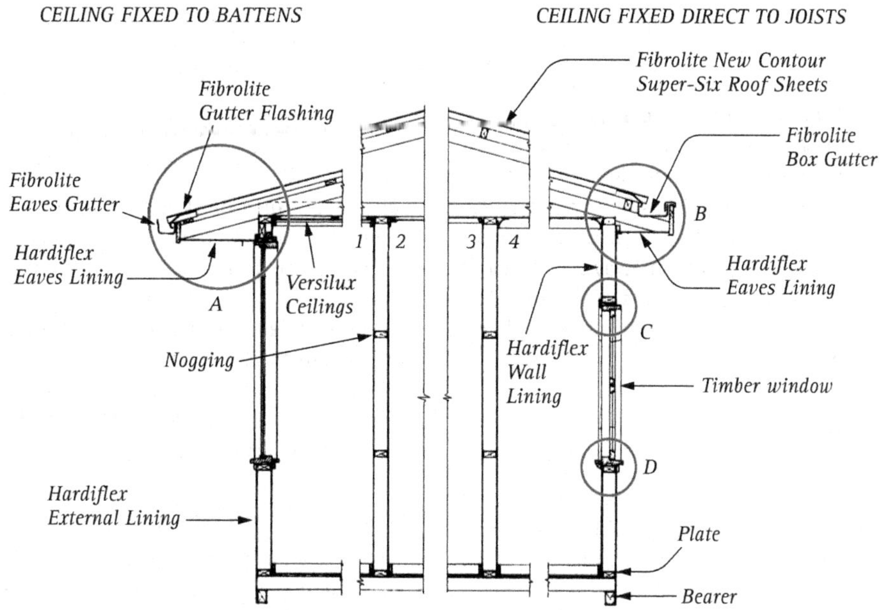

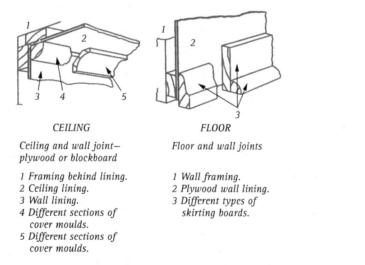

CEILING FIXED TO BATTENS

CEILING FIXED DIRECT TO JOISTS

Fibrolite New Contour
Super-Six Roof Sheets

Fibrolite
Gutter Flashing

Fibrolite
Box Gutter

Fibrolite
Eaves Gutter

Hardiflex
Eaves Lining

B

Hardiflex
Eaves Lining

1 2   3 4

Versilux
Ceilings

A

Hardiflex
Wall
Lining

C

Nogging

Timber window

D

Hardiflex
External Lining

Plate

Bearer

Figure 3.63 Section through typical timber-framed house

CEILING

*Ceiling and wall joint—
plywood or blockboard*

1 Framing behind lining.
2 Ceiling lining.
3 Wall lining.
*4 Different sections of
   cover moulds.*
*5 Different sections of
   cover moulds.*

FLOOR

Floor and wall joints

1 Wall framing.
2 Plywood wall lining.
*3 Different types of
   skirting boards.*

Figure 3.64 Details of finishing at junction of wall with ceiling and floor

with saw cuts to a wide piece of timber and use this like a mitre-box to cut studs to length without any marking. Do not measure each one with a rule as inaccuracies will occur. Select pairs of straight studs for door and window openings. Set out and cut

trenches for door heads and windowsills and heads.

Assembly

Assemble frames on the floor. Double-nail studs at top and bottom through plates. Cut and nail heads, sills and short studs in position. As each frame is assembled, square it, secure with a temporary brace, then cut in and nail the permanent braces. Where possible use two braces, one at each end from the top to the outer end of a frame (Fig. 3.62). Braces may be 50 × 25 mm battens let into the studs or they may be of light angle or hoop iron. Nogging, which is placed midway in the height of studs or at horizontal joints of the wall covering, may be cut in at this stage or when frames are erected.

Construction

When all frames have been assembled they are erected in turn, plumbed, temporarily braced in position, then the scarf at each corner is nailed. For a timber floor, bottom plates are nailed to the joists; for concrete, galvanised bolts in the concrete or concrete nails secure the bottom plates.

If the roof is of the skillion type, the frame assembly on the sloping ends requires different treatment at the top—a rafter becomes the top plate and studs are scarfed on to it. Bottom plates are the same as for the other walls.

One of the problems of frame construction is that not all the timber is straight. A solution is to select the straight pieces for long lengths, keeping the others for short lengths. If rounded pieces have to be used, face all rounded edges the same way, usually outwards.

ROOF FRAMING

A skillion or gable roof is normally used for a small building such as a garage or workshop. Skillion roofs (Fig. 3.65) are simpler than pitched roofs and need only a slight fall but, except in certain settings, are less attractive in appearance than pitched roofs.

Setting out a skillion roof

To set out a skillion roof, place a piece of timber in position for a rafter and mark the notching

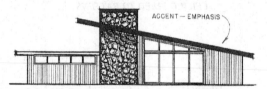

Figure 3.65 Shed or skillion roof is simple to construct and is attractive in certain settings, but requires extra insulation

and length at each end. Cut this rafter, place it in position to check size and fit (it is possible for the building to be out of parallel) and, when fit is correct, use it as a pattern, or template, for marking the others (Fig. 3.39).

Setting out a pitched roof

Once the span and rise, or pitch of a roof are known, all angles and lengths of members can be obtained. 'Span' is the overall distance between the outside of the wall plates. 'Rise' is the vertical distance the roof rises above the plates. 'Pitch' is the slope of the roof. 'Run' is half the span (Fig. 3.66).

The pitch of a roof may be expressed in three ways:

1. *as a pitch ratio (e.g. 1:2.84);*

2. *in degrees representing the angle formed by the hypotenuse and base of the triangle;*

3. *as a fraction of the span (e.g. ¼, ⅓, ½).*

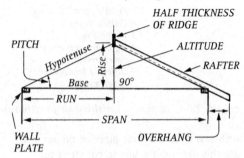

NOTE POSITION OF RAFTER COMPARED TO TRIANGLE ¼ PITCH

Figure 3.66 Right-angled triangles formed by members of pitched roof

To obtain the fraction, divide the rise by the span; thus 3000 mm rise and 9000 mm span give $\frac{3000}{9000} = \frac{1}{3}$. Remember that the span has two slopes of roof running from the central ridge, so that in setting out, 'A pitch becomes 333 mm rise in 500 mm of run. Roofs are seldom made to an exact fractional pitch but rather to a certain angle, or amount of rise per half metre, e.g. 300 per 500 mm. For a tile-covered roof, 30° pitch is the minimum for a good watertight job. This pitch works out at 290 mm rise in 500 mm run, which is between $\frac{1}{4}$ and $\frac{1}{3}$.

For most roofs the ridge is in the centre, so both triangles are equal in shape and size. Either of these triangles can be used to obtain the length and the angles of the plumb cut and foot cut of common rafters. In determining the length, remember that the top edge of all rafters is above the line of the hypotenuse.

Allowance must also be made for half the thickness of the ridge and hip rafter (Fig. 3.66). The right-angled triangle can be used for setting out in several ways:

1. *with the steel square (Fig. 3.67);*

2. *with a triangular piece of plywood or hardboard (Fig. 3.68);*

3. *by means of a drawing;*

4. *mathematically by the use of trigonometry.*

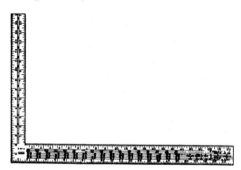

Figure 3.67 Steel square for use in setting out roof

TEMPLATE FOR COMMON RAFTER — LENGTH FOR 500 mm RUN — A — 250 — 500

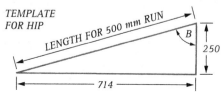

TEMPLATE FOR HIP — LENGTH FOR 500 mm RUN — B — 250 — 714

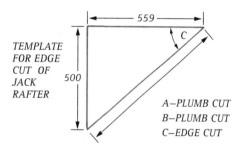

559 — C — TEMPLATE FOR EDGE CUT OF JACK RAFTER — 500

A–PLUMB CUT
B–PLUMB CUT
C–EDGE CUT

Figure 3.68 Triangles of plywood or hardboard are used similarly to steel square for setting out roof

All four methods are basically the same, though the application of the basic principle differs and methods 1 and 2 are the simplest and most frequently used.

Setting out with the steel square

Taking 500 mm as the base of the triangle, determine from the plan the amount of rise in 500 mm of the roof (Fig. 3.66). So that actual sizes can be stated in the following calculations, it is assumed that a quarter pitch, 250 in 500 mm, roof is being set out.

Length of common rafter. Place the steel square on a piece of straight timber to be used as a rafter, as shown in Figure 3.69. This gives the length of rafter for 500 mm run, which for a 250 in 500 mm roof is 559 mm. The total length can be obtained either by multiplying

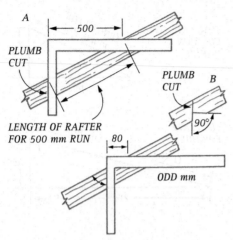

Figure 3.69 Using steel square to determine plumb cut and length of common rafter.

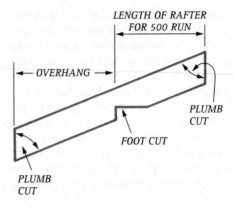

Figure 3.70 Template for foot of common rafter

559 mm by the total run or by stepping the distance off. It is a good plan to use both methods as a check. To obtain the length of rafter for odd millimetres see Figure 3.69. Make allowance for half the thickness of the ridge at the top and for overhang at the foot. Note that the top edge of the rafter is above the hypotenuse of the triangle (Fig. 3.66)

Angles of common rafters. The position of the steel square in Figure 3.69 gives the angles as well as the length of the rafter, 'A' being the plumb cut where the rafter joins the ridge and 'B' where the rafter fits on the wall plate. The cut across the edge is square. Set a bevel to these angles or, better still, make a template for them and for the overhang (Fig. 3.70). Complete the setting out for one rafter.

Cutting rafters. Saw one rafter to length and shape and use it as a pattern to mark out and cut another rafter. Place this pair in position on the roof and check for fit. Make necessary adjustments and use one of these rafters as a pattern for marking the others. The overhang may be cut off at this stage or later when the roof is pitched—the latter method ensures a straight line for the fascia and gutter. If the overhang is not to be enclosed, the feet of all rafters must be dressed.

Hipped rafters. For a hipped roof two more triangles are used to obtain lengths and angles, one for the hip (Fig. 3.71) and one for the jack or creeper rafters (Fig. 3.72).

The length and angles for the hip are obtained by using the triangle 'A'. The hip is at 45° to the side and end of the building, so it is a diagonal of the run. The diagonal of 500 mm and 500 mm is 714 mm, so the triangle has a 714 mm base and a 250 mm altitude (Fig. 3.71). This gives a length of 500 mm run of the hips as 750 mm. The plumb and foot cut angles are 250 and 714 mm (Fig. 3.71).

When setting out jack rafters, the length for each 500 mm run, the plumb cut and foot cut

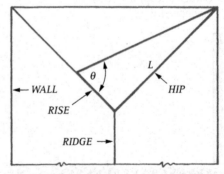

θ = PLUMB CUT FOR HIP, L = LENGTH OF HIP

Figure 3.71 Determining lengths and cuts for hip rafter

angles are the same as for the common rafters, but the angle for the edge cut, always less than 45°, is 500 to 559 mm (Figs 3.72, 3.73). Cut a template for this angle (Fig. 3.74). Jack rafters are cut in pairs, one pair for each hip. It is a good idea to mark out the length of all jack rafters on one of the common rafters.

Setting out with triangular shapes

Triangular shapes can be used to set out in practically the same way as with a steel square. Set out and cut the triangles to shape (Fig. 3.68): one for common rafters, 500 base to 250 mm altitude; one for the hip, 714 to 250 mm; and one for the edge cut of the jack

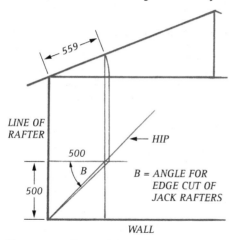

Figure 3.72 Determining edge cut for creeper, or jack, rafter

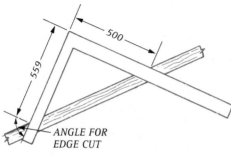

Figure 3.73 Using steel square to obtain edge cut for jack rafter in quarter-pitch roof

rafters, 559 to 500 mm. These triangles will give lengths and angles in the same way as the square.

Pitching the roof

The length of the ridge is easily obtained from the setting out on the wall plates. Cut the ridge to length and mark the positions of all rafters on it. Fix a pair of rafters, without the ridge, at each end of the gable section and then fix the ridge between them. Plumb up the ridge and fix a temporary brace to hold it in the correct position. Fix all common rafters in position.

Fix the hips to the end of the ridge and mitre the joints on the edges. Fix the centre pair of jack rafters to each hip, then the remainder. Take care not to make the hip crooked by overdriving one of the rafters—check the hip for straightness by eye or fix a chalk line above one of its edges.

Setting out valley rafters and jack rafters is similar to setting out for hips. Rafters may run from ridge to valley or hip to valley.

It is usual to single-nail all roof members during pitching and double-nail on completion.

Purlins are nailed on edge in the centre below rafters, then struts and collar ties (Figs 3.35, 3.36). Remember that struts must be fixed above a wall. In the absence of a centre wall, collar ties are used (Fig. 3.35).

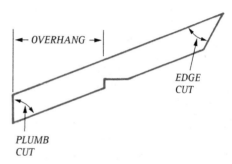

Figure 3.74 Template for foot or jack rafter

Battens of a size and spacing to suit the type of roof covering are fixed. (See 'Roof covering materials', pp. 74-5.) Methods of finishing eaves are shown in Figures 3.43 and 3.48.

WALL COVERINGS

The two most common wall-cladding materials are weatherboards and fibre cement. Also useable are waterproof plywood and tempered hardboard, the latter offering a special 10 mm thickness.

Weatherboards are nailed to the studs in a variety of section patterns (Figs 3.29, 3.57). Corners may be mitred, but the usual method is to fix a stop at each corner, cutting the weatherboard up to it (Figs 3.57, 3.60). Boards are cut flush with the inside of door and window openings. Weatherboards are made to overlap and need no flashing between boards, however, they should be primed before fixing.

Fibre cement is available in a number of section patterns (Fig 3.58) but plain flat is the easiest to fix. Horizontal butt joints must be flashed with galvanised iron, aluminium or lead (Fig 3.48), while vertical joints in plain fibre cement are finished with a cover strip or joint moulding (Fig 3.48). Patterned fibre cement has overlap at the joints. External and internal angles are covered with a special angle strip of aluminium or plastic.

Sheets should always be stacked flat on a smooth surface or on edge, with little slope, against a wall or similar support.

Plywood and hardboard are usually finished at vertical joints with a wood cover strip, but these joints may be filled with special waterproof fillers. External angles are covered with timber while internal angles are generally finished with quadrant moulding (Fig 3.64).

So that sheets remain flat after they are nailed to the wall, they should be wetted on the back and stacked flat, back-to-back, 24 hours before use.

Flashing

Window heads, door heads and sills must be flashed to prevent leaks (Fig 3.60). For door and window heads, the flashing goes under the wall covering and out over the architrave. For sills, the flashing goes under the sill, is turned up around the ends and on the inside of the sill and down on the outside of the wall covering. Sill flashing is covered on the inside and the outside with an apron piece under the nosing and below the sill.

Flashing may be of thin lead, 26-gauge galvanised iron or bitumen-coated aluminium. Lead is expensive but the most satisfactory, since it is long lasting and easily worked.

FRAMES FOR OPENINGS

The construction of timber frames is described here, but it should be noted that many buildings nowadays incorporate aluminium door and window frames as well as sashes.

Door frames

The doorframe to which a door is hung and secured requires a rebate into which the door fits to make a rainproof, windproof and lightproof joint. Rebates may be formed by nailing on a stop or bead but are preferably run in solid timber. Note that frames for sliding doors are not rebated and are not used to secure the door. Frames are braced after assembly, the braces being left in place until the frames are held securely in the wall. All doorframes must be fixed plumb or the door will tend to open or close when not fastened.

Frames for external doors in brick buildings, usually run from 75 mm-thick timber, are built in during construction of the walls. They are dowelled into the step and held in the cavity with wooden cleats or sometimes by hoop iron or wire mesh nailed to the frame and built into the brickwork. An ovolo or quadrant mould is nailed to the outside edge of the frame and the brickwork on the outside wall fitted to it. For timber buildings, external doorframes are usually thinner. They are dowelled into the step and nailed to door studs (Fig 3.75).

Internal door frames, called 'jambs', are either wide enough to allow architraves to be fixed to both edges over the wall covering, or narrow, requiring the use of a stronger mortar in the brickwork to prevent cracking which could result from the movement caused by the door slamming. Jambs are nailed into plugs in the reveal of the opening in brickwork or into studs in a timber building. They are finished with quadrant or ovolo moulding to each edge inside the brick reveal.

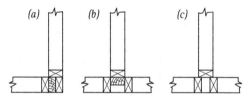

Fig 3.75 Wall studding at corners and internal walls:
(a) pairs of studs on weatherboard corners;
(b) 3 × 200 mm blocks, equally spaced;
(c) pairs of studs at T junctions

Door frames in a house must allow doors to open inwards to all rooms, and, in bedrooms, on the side that will not expose the whole room if the door is left partially open (Fig 3.76).

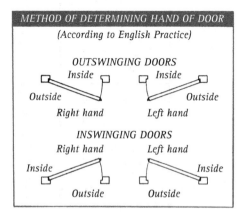

Fig 3.76 Doorframes must allow for doors to open in required direction

Standard size of house doors is 2040 mm high and 820 mm wide. Entrance doors may be wider and toilet and bathroom doors narrower, but all doors are the same height so that architrave heads will be in line.

Window frames

A window consists of a frame fixed in the wall opening and one or more sashes. The sash is the frame that holds the glass and is either fixed or movable. Sizes of windows are not standardised but unless the height extends to the ceiling, the heads are usually in line with door heads.

Window frames vary in section according to the kind of sash, of which there are five in general use: fixed, double-hung or box, casement, hopper and sliding. Fixed sashes are used only in conjunction with opening windows as they do not provide ventilation.

Box frames, for sashes which slide up and down with counterweights or balances, are built up. The outside lining, the parting bead and the stop bead form the two grooves in which the sashes slide. Older types of box frames were fitted with axle pulleys, cords and cast-iron weights running in a box (hence their name). Modern types have sash balances operated with a spring and do not require a box. Frames for hopper, sliding and casement windows are usually solid rebated, although stops may be nailed on to form a rebate.

Frames are built in during construction of brick walls, being held with cleats built into the cavity. In timber construction, frames may be placed in position at any suitable time, usually when wall coverings are being fixed.

All window frames require a sill, usually of hardwood, tallow wood being the most suitable. Sills must be stepped and properly throated to prevent rain driving underneath the sashes. Flashing must be provided beneath the sill, turned up on the inside edge of the sill and down into the cavity of a brick wall or on to the outside of a timber wall (Fig 3.60).

Parts of door and window frames

The parts of frames are named and defined as follows:

Stiles—*The vertical members of frames and also of doors and sashes. The hinged side is called the 'hanging stile' and the lock side the 'closing stile.'*

Head—*The top horizontal member of a frame.*

Step—*The bottom horizontal member of a doorframe, not generally built in with the frame.*

Sill—*The bottom horizontal member of a window frame.*

Mullion—*The vertical member of a frame between head and sill.*

Stop bead—*Nailed to a frame to form a rebate or groove.*

Box frames include other members, namely:

Outside linings—*Project about 16 mm beyond the inside of the stiles and head to form one part of the groove in which the sash runs.*

Inside linings—*Finished flush with the inside of the stiles on which the stop beads are fixed.*

Parting bead—*Fixed into a groove in the stile to part two sashes and form a groove in which the sash slides.*

DOORS

Doors are of five main types:

1. *flush doors having a centre frame with plywood or hardboard glued to both sides;*

2. *framed doors having a frame either with panels or with a facing of T & G lining boards (Fig 3.77);*

3. *ledged doors with horizontal ledges to keep them square (Fig 3.78);*

4. *framed and sheeted doors (Fig. 3.79);*

5. *roller doors made of wood or metal interlocking plates which enable the door to be rolled up.*

Butt or T hinges are used for hanging doors and there are special slides and supports for sliding doors (Fig. 3.80). Roller doors are rolled on to a central roller operated by gears and chains. Garage doors may be hung on balanced supports so that the door opens by tilting upwards to a horizontal position. Locks are of rim or mortice pattern and special locks

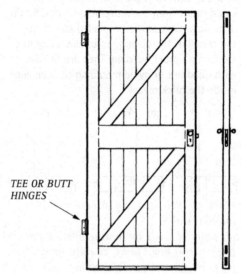

Figure 3.77 Details of framed, ledged and braced door

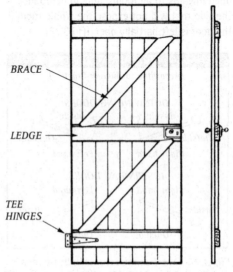

Figure 3.78 Details of ledged and braced door

and fasteners are supplied for sliding, roller or tilted doors (Figs 3.81, 3.82, 3.83).

Construction of garage doors

In addition to the large main-entrance doors, a garage may be provided with a single side door of similar construction. Doors for garages can be ledged and braced (Fig. 3.78), framed, ledged and braced (Fig. 3.77), or framed and sheeted (Fig. 3.79).

For the main entrance, the first two types must be installed in pairs and are less pleasing in appearance than the sheeted door, which is usually faced with waterproof plywood and can be made to cover the whole opening. Also their weight creates problems in making and hanging as well as causing hinge wear and a tendency to sag. However, a disadvantage of the ply-sheeted door is that unless it is kept painted, particularly round the edges, water soaks in and causes the ply to buckle and split.

Ledged and braced doors. (See Fig. 3.78.) The three ledges of 150 × 25 mm dressed timber with edges bevelled or chamfered are placed about 150 mm from each end with the one called the 'lock' in the middle. The covering is of T & G V-jointed timber with a finished thickness of at least 16 mm.

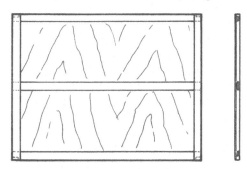

Figure 3.79 Details of framed and sheeted door

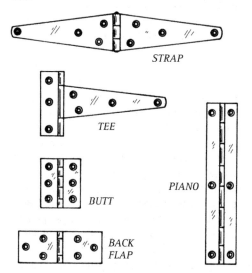

Figure 3.80 Typical hinges

STRAP

TEE

BUTT

PIANO

BACK FLAP

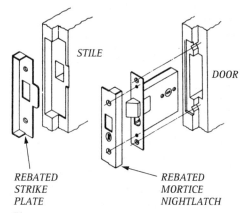

STILE

DOOR

REBATED STRIKE PLATE

REBATED MORTICE NIGHTLATCH

Figure 3.81 Installation of rebated mortice lock

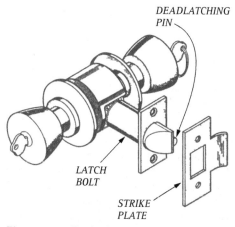

DEADLATCHING PIN

LATCH BOLT

STRIKE PLATE

Figure 3.82 Typical mortice lock with key, face plate and striking plate

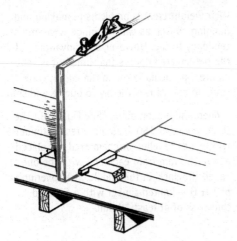

Figure 3.83 Block and wedge holding door for edge planing

Braces of 75 × 25 mm are notched into the ledges with the lower end of each brace toward the hanging edge of the door.

Framed, ledged and braced doors. (See Fig. 3.77.) The frame consists of stiles, rails and braces. The stiles and top rail are not less than 75 × 50 mm DAR (dressed all round), and the other rails are thinner so that the lining boards finish flush on the face of the stiles and top rail. Use bare-faced mortice and tenon joints on the thinner rails and haunched mortice and tenon on the top rail.

Ideally, lining boards for ledged doors are all the same width but the two outside boards may have to be made narrower to achieve the desired width of the door. Since boards may vary slightly in width due to shrinkage or uneven milling, check the overall width before making the frames, remembering to allow for removal of the tongue and groove respectively from the two outside boards.

Framed and sheeted doors. (See Fig. 3.79.) Since the waterproof plywood covering is much lighter in weight than boarding, the frame for these doors can be much smaller in section than for the framed ledge door and bracing is unnecessary. The covering is usually attached with a waterproof glue rather than rebated in.

Framed and sheeted doors are the type used with the tilting system of hinging.

Fitting and hanging doors

Since they are large and exposed to the weather, garage doors are fitted with not less than 3 mm clearance. Pairs of doors may be left plain on the closing side, or solid-rebated, or fitted with a bead to form a rebate.

To fit a framed door, saw the bottom horns from the stiles, prop the door against the opening and run a pencil mark all around the inside. Saw off the top horns and plane to the lines all round. Figure 3.83 shows a block and wedge holding the door for edge planing.

Doors may be hung with butt hinges or, preferably, with galvanised T or back-flap hinges, in which the pull on the screws is not direct (Fig. 3.80). To avoid undue leverage on the hinges and slackness in the hang of the door, hinges should be as close as possible to the top and bottom. T and back-flap types should be fixed to the outside and located so that screws will penetrate the rails.

Butt hinges may be plain or similar to a T hinge with an ornamental plate. They should be of the wide, heavy type so that screws will not come too near the front of the frame, and must be located clear of the rails to avoid screwing into the joint. Hinges may be positioned either by fitting the door in the opening and marking both door and frame or by fixing the hinges to the door and then fitting it in the opening to mark the frame. Figure 3.84 shows three ways of recessing and positioning butt hinges. In all cases accuracy in setting out and cutting is essential. Mark the length of the recess with a pencil and gauge the depth and width, preferably using two marking gauges as shown in Figure 3.85. On rebated doorframes, use a butt gauge (Fig. 3.86) which has two spurs positioned to allow for clearance between the inside of the door and the back of the rebate. Figure 3.87 shows how the recess is cut and pared out. Secure the hinges to the door and recesses with one centre screw placed so that it will draw the hinge back to the inner edge of the recess.

Figure 3.84 Butt hinges may be fixed in three positions; note variations in recessing

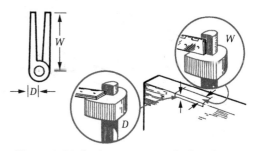

Figure 3.85 Setting out recess for butt hinge

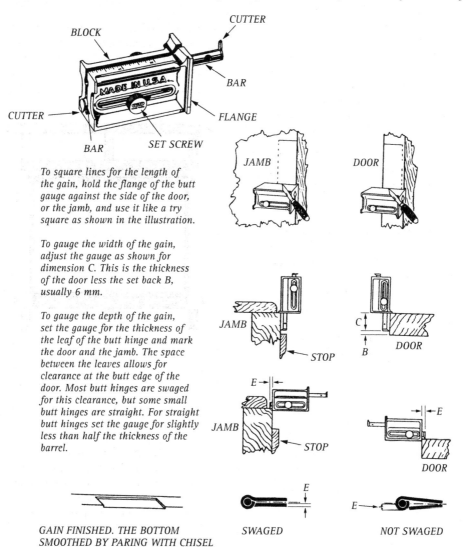

To square lines for the length of the gain, hold the flange of the butt gauge against the side of the door, or the jamb, and use it like a try square as shown in the illustration.

To gauge the width of the gain, adjust the gauge as shown for dimension C. This is the thickness of the door less the set back B, usually 6 mm.

To gauge the depth of the gain, set the gauge for the thickness of the leaf of the butt hinge and mark the door and the jamb. The space between the leaves allows for clearance at the butt edge of the door. Most butt hinges are swaged for this clearance, but some small butt hinges are straight. For straight butt hinges set the gauge for slightly less than half the thickness of the barrel.

GAIN FINISHED. THE BOTTOM SMOOTHED BY PARING WITH CHISEL

SWAGED

NOT SWAGED

Figure 3.86 Butt gauge for setting out hinge recess on rebated door frame

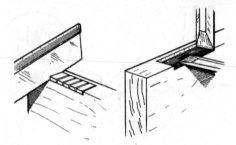

*Figure 3.87 Cutting recess for butt hinge;
accuracy is essential*

Place the door square in the opening,
supporting it on a long, thin wedge or a thick
chisel, which can be adjusted to allow the
hinges to be fitted into the recesses on the
jamb. Secure them with a centre screw and
check for correct hang (both doors in the case
of a pair) before driving the other screws.

Pad bolts at top and bottom hold one door of
a pair and the other is fastened to it by a
similar bolt, or a hasp and staple (Fig. 3.88); a
rebated mortice lock (Figs 3.81, 3.82) or a
night latch makes a better-looking fastening.
Figure 3.81 illustrates installation of a mortice
lock. To fit a night latch, bore a hole for the
barrel, sized and positioned according to the
instructions supplied with the lock. Cut screws
and tongue of barrel to length to suit the
thickness of the door. Screw the barrel into
place, nail the back plate in place and screw
the latch over the tongue. Position the box of
the lock with the door closed and screw the
box into place.

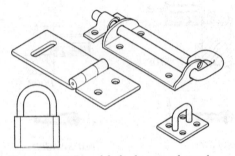

*Figure 3.88 Barrel bolt, hasp and staple
and padlock*

ALUMINIUM WINDOWS

The majority of windows used in house or
cottage construction are manufactured from
aluminium, anodised and powder coated for
maintenance-free service.

When ordering new windows double check the
rough stud opening measurements and request
frames and glass to be made in accordance
with Australian Standards. Request reveals and
flashings are factory fitted.

Flashing of sheet lead or other material should
be provided over the head, underneath the sill
and on each side of the window. The flashing
is fitted into the opening before the window is
placed in position from the outside. (See Fig.
3.89.)

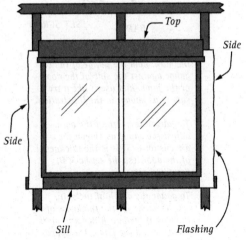

*Figure 3.89 Flashing required for aluminium
windows. Fix sill, sides then top in that order*

Sill flashing

The sill flashing needs to be folded to form a
one-piece tray with a 5mm turn up at the
ends and be housed into the jamb stud. The
turn up of the back of the tray should be at
least 25 mm and the turn down over the face
of the external wall sheeting or covering also
25 mm. Care must be taken to fold the
flashing in order to form a watertight tray.
(See Fig. 3.90.)

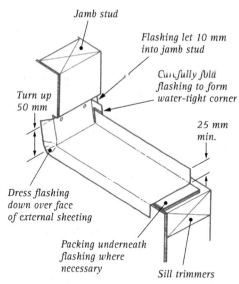

Figure 3.90 Method of folding flashing under windowsill to form a watertight tray

Head flashing

The head flashing is turned up under the wall sheeting and then turned down over the frame or architrave (see Fig. 3.91). A space of at least 12 mm should be left between the top of the window frame and the lintel to allow for settlement of the building frame.

Head flashing is unnecessary if the top of the window frame is close to the overhanging eaves.

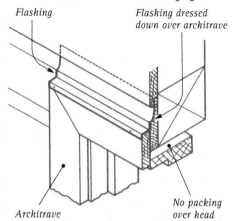

Figure 3.91 Head flashing should fit under weatherboard or fibreboard sheeting.

FIXINGS

The various members that finish a room at floor, doors, windows and corners are known as 'fixings'.

Architraves around door and window openings cover the junction of frame and wall and also help secure the frame. They are provided around the outside and inside of external doors and windows in timber buildings but on the inside only in brick buildings where elbow linings are nailed to the frame to make up the difference in width between frames and walls. Consisting of a head and two vertical members called 'stiles', and usually mitred at corners, architraves are generally of simple sectional shape in 75 × 25 mm or 100 × 25 mm timber. At windows they rest on the sill nosing, below which an apron piece is fitted.

Skirting is the timber nailed around the bottom of the wall at the floor. Modern trends are for narrow skirtings (Fig. 3.64).

Picture rails, if installed, are usually positioned at head height. They are smaller than architraves but have the same sectional shape. The top inside corner is backed off to allow for picture hooks.

Corner finishes are of many types, depending on whether they are for internal or external corners. External and internal angles on the outside of a timber building require a stop for weatherboards (Fig. 3.57) and angle coverings for fibre cement (Fig. 3.58). On the inside of a timber building, the angles are either set or jointed with perforated metal angle strips.

Beading is a term often incorrectly used to describe different types of mouldings in general whereas beads are specially shaped—half-round on their edges and all except the parting beads have a quirk, or channel. Beads are generally used as some form of stop or spacing piece. Most window frames have stop beads while box frames have both parting and stop beads. Cock beads are used on a face or between butt joints. Beads are also used to make joints less noticeable, as in lining boards.

The moulding covering the edges of the wire mesh on fly proof doors and window screens is often miscalled 'beads' or 'beading'. In fact, it is 'half-round moulding', often called 'fly mould' because it is used so often on fly screens. It is also applied over the butt joints of fibre cement, hardboard or plywood.

Corner joints of fixings and beads are mitred or scribed. In a brick building the wall must be plugged before plastering to take all fixings and doorjambs.

FITMENTS

Built-in cupboards may be constructed to form a partition between rooms, with one section opening into each room. They may be from floor to ceiling, with sliding or swinging doors the same height as main doors, and sliding doors on the top portion. Wardrobe cupboards should be at least 500 mm deep to allow clothes to be hung sideways, but storage cupboards, usually fitted with shelves, may be made to any suitable depth.

Butt hinges are used to hang doors, while special slides with quiet runners are used for sliding doors. Locks of the rim or mortice patterns and fasteners of the clip or magnetic type may be fitted. Handles are decorative and can be matched to the furniture handles in the room.

In kitchens, floor cupboards equipped with shelves and drawers usually reach to sink height. They may include the stove as part of the unit, with an oven built into the wall. Cupboard tops covered with plastic laminate form a working bench. Wall cupboards are fixed to the wall above the floor cupboards with a working space between.

HINGES AND THEIR USES

The days of cabinetmakers making their own hinges are long since gone, from a practical point of view. Today there is a tremendous range of hinges available, and it is essential to select the correct hinge for a particular job.

Most hinges are designed to be as inconspicuous as possible when fitted. However decorative hinges are meant to be displayed and are mounted on the outside of a door or carcass. Butt hinges, for example, have to be recessed into both the door and cabinet so that the door fits tightly when closed, with no unsightly gap. The flaps of the hinge can be recessed into the door edge and carcass or one flap may be fully recessed into the door. Alternatively, there is a flush hinge which does not need to be recessed, as the two flaps interlock when closed.

Concealed hinges can be hidden by the door when it is closed or neatly inset into the edges of the door and carcass.

Hinges are made in a variety of materials. The best for cabinet work are made from extruded brass which are superior to those of folded brass. Decorative hinges are available in a variety of finishes with the surface protected by lacquer. It is important not to polish decorative hinges as they will discolour. Many hinges are also made from steel and plastic.

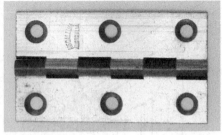

(a) Butt hinge

This hinge is suitable for all doors. It is common practice to recess the two flaps so that the door fits tightly against or within the carcass so that only half of the knuckle is visible. The basic type is made of steel, usually with an electroplated brass, or japanned finish.

(b) Back-flap

This is a standard hinge for table leaves and fall-flaps on cabinets and desks. It can be either surface mounted or recessed into the surface of the carcass and the flap and it is fastened to the underside of the wood. The wide flaps provide strong support for the weight on the table leaf.

(c) Flush hinge

This hinge is made of thin coated metal and has one flap which rests inside the other when the hinge is fully closed. It is essentially a lightweight hinge that is best used on small cabinet doors. It does not have to be recessed and can be screwed directly to both door and carcass.

(d) and (e) Concealed cabinet hinge

A concealed hinge suitable for lay-on doors. The degree of swing means it can be open without disturbing the doors on either side

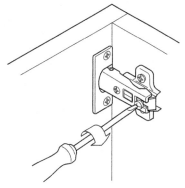

of it. The section linked to the door has to be recessed, but the arm is surface mounted inside the carcass. These hinges are normally spring-loaded.

(f) Soss hinge

This is the original invisible hinge and is for use with lightweight doors. It consists of a number of interconnecting scissor joints. These help to distribute the load of the door on the hinge to produce a smoother opening action. The angle of opening is 180 degrees.

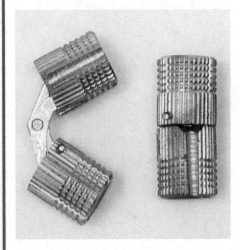

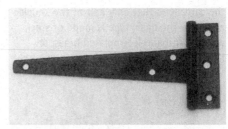

(i) Tee hinge

An extra long flap makes this hinge suited for gate and ladder construction. Normally made of steel, black coated or galvanised for exterior use.

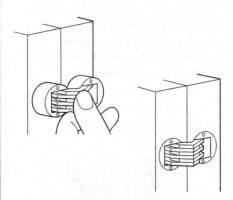

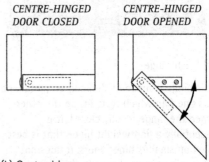

(j) Piano hinge

A continuous form of butt hinge used where greater strength is required. Used for swinging doors in kitchen cabinets, hinged desk flaps, and piano lids.

| CENTRE-HINGED DOOR CLOSED | CENTRE-HINGED DOOR OPENED |

(g) and (h) Cylinder hinge

A further development of the soss hinge suitable for fall flaps, flush tops, and cabinet folding doors. It is simply set into holes drilled in the edges of the door and cabinet and is completely concealed when the door is closed. Small adjusting screws allow for a tight fit. The angle of opening is 180 degrees.

(k) Centre hinge

Sometimes referred to as a single-cranked hinge for lay-on flush fitting cupboard and cabinet doors. Usually the flap is screwed directly to the carcass, without being recessed, and the crank goes around the end of the door.

KNOCK DOWN FITTINGS AND THEIR USES

Knock-down fittings have revolutionised furniture construction. Instead of structural joints a particular reliance has been placed on knock-down fittings, which allow for simple assembly and dismantling. Some are designed for furniture, which is made from manufactured board, others for solid timber frames.

Cost is reduced as they avoid the need to cut joints, often calling only for the boring of one or two holes, instead of a more complex process. They eliminate gluing up and factory assembly altogether.

In general, they are unsuitable for traditional pieces of furniture, but their convenience makes them otherwise useful. They all require accurately drilled holes.

the core of the thread is required. The head can be covered with a plastic cap if needed

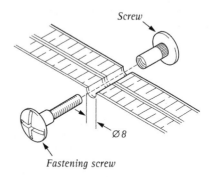

Fastening screw

(c) Cabinet connector

Used for linking prefabricated units together against a wall, for example in the kitchen. Made of plastic, it is designed for use with standard particleboard thicknesses.

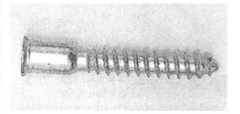

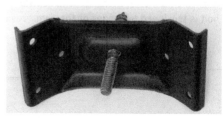

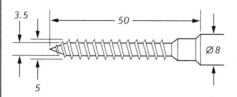

(a) and (b) Screw connector

Sometimes called a chipboard screw, this is a blunt-ended steel fastener with a wide, shallow thread, which gives it a holding power superior to that of even the most sophisticated conventional screw. A pilot hole the same diameter as

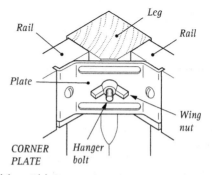

(d) and (e) Corner plate

For a simple and rational connection of table frames and legs. The fitting is

inserted in a milled aperture and secured with screws. The table legs are screwed on through the middle hole. Tables fitted with these corner braces can be dismantled for storage, packaging and transportation.

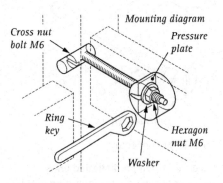

Mounting diagram

Cross nut bolt M6

Pressure plate

Ring key

Hexagon nut M6

Washer

(f) Cam connector

A special purpose fitting for connecting lengths of plastic laminated particleboard kitchen bench tops, desks etc. The central bolt is tensioned by means of the cam.

(g) and (h) Bolt and barrel nut

This fixing is so strong that it is used to join table and chair frames. The bolt

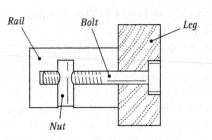

Rail   *Bolt*   *Leg*

Nut

BOLT AND BARREL NUT

passes through upright or leg into the rail, where it meets a threaded hole in the barrel nut. Wooden locating dowels in the end grail keep the rail aligned correctly as the bolt is tightened up with an Allen key.

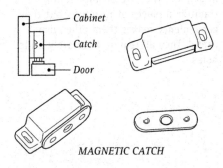

Cabinet

Catch

Door

MAGNETIC CATCH

(i) Magnetic catches

Small encased magnets are screwed to the inside of the carcass side panel or located in a hole drilled in the edge. The magnet attracts a flat metal striker plate fixed to the cupboard door.

PAINTING

Advances in the paint industry are rapid and new products are continually being introduced. Do not assume that all paints are the same, but study the free technical advice offered by manufacturers, who have tested their products and know the procedures required to achieve a good result. All stress the importance of correct surface preparation,

correct choice of primer or undercoat, care in painting end-grain of timber, stopping of all cracks and holes, rubbing down and allowing the specified drying time between coats and use of the appropriate type of thinners.

Primers, typically pink or white, are designed to prepare the surface of new wood for the application of finishing coats. They are slow drying and therefore have good binding and penetrating qualities, in contrast to finishing paints, which do not penetrate because of their fast-setting properties. New timber should always be primed and unless the correct primers are used finishing coats are likely to crack and peel. For metal, zinc chromate primer is recommended as giving good adhesion for finishing coats.

Sealers are designed to overcome porosity and are used mainly on absorbent materials like fibre cement and fibrous plaster to prevent finishing coats drying with an uneven appearance. However, flat or satin finishes may be applied to fibre cement or wallboards without undercoating or sealing.

Binders are used on previously painted surfaces for binding water-soluble paints which cannot be completely removed. Usually binders act also as sealers.

Stopping is for filling small cracks, knot holes and nail holes. On new work it should be applied after priming since adhesion is unsatisfactory on bare wood and the stopping may crack and fall out. Stopping for general woodwork is putty composed of fine whiting mixed with raw linseed oil to a soft, even consistency. Unused putty is kept soft by placing it in a tin or jar and covering with water. Other types of putty, not necessarily the whiting and linseed oil mixture, are available for specialised work. Examples are glazing, quick-setting, metal, aquarium and sanding putties. Stopping specially suited to the treatment of building boards, such as plasterboard and fibreboard, is marketed under various trade names. Mixed with paint to make what is known as 'Swedish putty', it can be used on woodwork. Both ordinary and Swedish putty may be coloured with paint or paint stainers to match finishing colours.

Strippers for removing old paint before repainting require careful use. They are made to dissolve paint and, being in liquid form, they penetrate into the wood. Subsequent coats of paint may be affected, so most strippers must be

neutralised before repainting is commenced. The manufacturer's instructions should be followed.

Preparation of surfaces

It is important to prepare surfaces thoroughly if a good finish is to result. Remove the arris from all corners and clean up surfaces such as door and sash faces. If a clear paint is to be used, the surface must be carefully sanded with the grain—cross-sanding will show scratches in finished work. Dust down all surfaces before any painting is commenced.

Applying paint

Paint may be applied by brush or roller. Brushes are best for narrow surfaces and trim, such as architraves, but a narrow roller is fast and effective on such work as panelled doors. Manufacturers recommend different types of rollers for different types of paints and their advice should be followed. A soft, long-bristled brush produces a better finish with most paints. Choose a 38 or 50 mm brush for sashes and similar narrow work, and a 75, 90 or 100 mm brush for wider surfaces.

Begin at the top of the work and move from right to left, flowing the paint on and brushing out to an even surface so as to avoid 'runs'. Do not dip the brush too deeply in the paint, and wipe off the surplus on a wire stretched across the tin. Acrylic paints dry very quickly and to keep them flowing easily the brush should be dipped in water at frequent intervals. In hot weather, the surface may be dampened before painting.

For a good finish on woodwork, a priming and one other coat are needed before the final coat. Allow the primer to dry thoroughly (24 hours), apply stopping with a putty knife, pressing it firmly in to fill the holes so that there is no risk of sinkage, remove surplus with the knife, sand whole surface smooth and dust down with a dry brush before applying the next coat. Check the stopping, sand lightly, and dust again before the final coat. Time is saved if parts such as door and window frames, doors and sashes, and rafter overhangs are prime-coated before fixing.

Colour selection

Skilful use of colour will improve the appearance of any structure. The modern trend is to use light colour for large surfaces and contrasting trim on gutters and architraves, with doors painted in solid, bright colour or in panels picked out as a feature. Light colours tend to make a building appear larger than it is and dark colours smaller. Horizontal lines accentuate length and vertical lines height. Most brands of paint offer matching colours in flat, satin and gloss finishes.

Care of brushes and rollers

Paint should never be allowed to dry in the brush or roller. Between coats or overnight, a brush for water-thinned paint should be left suspended in a tin of water by a strong wire passed through a hole bored in the brush handle.

Thorough cleaning is essential after completion of painting or at change of colour. Brush or roll out as much paint as possible. A roller is quickly cleaned by drawing a loop of thick wire along it. Wash thoroughly in mineral turps, water or thinners, according to the type of paint, until no colour shows in the cleaning liquid. Final washing in warm water and mild soap or detergent followed by rinsing and drying is recommended for equipment that will be stored and not used again for some time.

OCCUPATIONAL HEALTH AND SAFETY

Occupational Health and Safety (OHS) is a term used in Australia to describe a number of workplace practices combined with a number of state and federal laws that are aimed at improving the standards of workplace health and safety. Workers in the building and construction industry must be aware of the occupational health and safety requirements of their industry as well as the responsibilities of employers and employees.

Historically the building and construction industry has one of the highest incidents of work-related injury or illness. Thousands of people each day suffer from some type of work-related injury or illness. The prevention of accidents is the concern of all people involved in the industry; all workers must learn to how to work without hurting themselves or their fellow workers.

OHS LEGISLATION

Each state in Australia has had to introduce laws that improve workplace safety practices. This has been overseen by the federal regulatory authority called Worksafe Australia under the *National Occupational Health and Safety Act 1985*.

This legislation was brought about due to the growing cost of compensation cases due to work-related accidents and injuries as well as the inequity of only approximately one third of the work force actually covered by the existing OHS laws.

There was the need for a standardisation of OHS from state to state. Now each state in Australia has a similar Occupational Health and Safety Act that aims to protect the health, safety and welfare of people at work.

STATE REGULATIONS

Each state now has its own regulations that must be met. These should include the following.

Employers

The employers must provide for the health, safety and welfare of their employees at work.

Employers must:
- Provide and maintain equipment and systems of work that are safe and without risk to health;

- Make arrangements to ensure the safe use, handling, storage and transport of equipment and substances;
- Provide the information, instruction, training and supervision necessary to ensure the health and safety of employees at work;
- Maintain places of work under their control in a safe condition and provide and maintain safe entrances and exits;
- Make available adequate information about research and relevant tests of substances used at the place of work.

Employers must not require employees to pay for anything done or provided to meet specific requirements made under the Acts or associated legislation. They must also ensure the health and safety of people visiting their places of work that are not employees.

The Employees

The employees must take reasonable care of the health and safety of others. They must co-operate with employers in their efforts to comply with OHS requirements.

Employees must not:
- Interfere with or misuse any item provided for the health and safety or welfare of persons at work;
- Obstruct attempts to give aid, or attempts to prevent a serious risk to the health and safety of a person at work;
- Refuse a reasonable request to assist in giving aid or preventing a risk to health and safety.

Workplace Hazards

The building industry has many hazards that may lead to death or serious injury. Things such as electricity, falls, collapsing trenches and melanoma often cause death. Noise, dust and chemicals can result in burns, blindness, deafness and injuries to the lungs. Back injuries or other serious strains can put workers out of work for lengthy periods of time, if not permanently.

These hazards are things that are potential sources of harm to anyone who is exposed to them, things that may affect the health and safety of a person. If every person in the industry is aware of the hazards to health and safety that they may be exposed to, and follows commonsense safety rules, both employer and employees will benefit.

We must be able to identify workplace hazards at work. Some of the common hazards that influence health and safety in the building industry are:
- Noise
- Fire
- Visibility (lighting)
- Vibration
- Falls, both objects and people
- Lifting and handling materials
- Machinery.

COMMON WORKPLACE HAZARDS ON BUILDING SITES

Safety Hazards

Poor housekeeping—untidy sites, lack of guardrails, inadequate walkways and rubbish left in work areas.
- Handtools—if not used and maintained correctly.

- Electrical tools—unguarded moving parts, electrocution, or falls and burns from electrical shock.
- Electricity—underground and overhead supply cables, site wiring and power leads.
- Water—slips and falls, electrocution, drowning and caving in of excavation works.

- Ladders and scaffolding—slips and falls from ladders and scaffolding.
- Heavy equipment—heavy motorised equipment moving around construction sites is a major hazard.
- Sharp materials—sharp edges can cause cuts and lacerations.
- Airborne materials (projectiles)—falling or flying materials.

Physical Hazards

- Noise—hearing loss, stress and annoyance.
- Heat and cold—cause reduction in concentration and heat-related medical conditions.
- Vibration—whole-body or hand and arm vibration injuries caused by the use of tools which vibrate.
- Manual handling—lifting, carrying, pushing, pulling.

Chemical hazards

Many different chemicals that may be harmful to health are used in the construction industry in the forms of:
- Solids—dusts, fumes and solid materials.
- Liquids—liquid materials and mists.
- Gases—gases and vapours.

Some of these chemicals can cause acute or chronic injuries or medical conditions.

Biological hazards

- Poor sanitation in dining areas, toilet areas and poor practices of rubbish disposal increase the chances of the spread of disease.
- Micro-organisms—Some work areas may also harbour dangerous bacteria.

Stress hazards

Stress is normally experienced as fatigue, anxiety and depression.

HAZARD CONTROL PROCEDURES

In many cases a number of control methods are used to control a hazard. Various methods should be considered.

The following list emphasises the control of the hazard at the source.

Engineering controls

- Design—try to ensure that hazards are 'designed out' when new materials, equipment and work systems are being planned for the workplace.
- Enclose or isolate the process—e.g. through the use of guards or remote-handling techniques.
- Alterations to tools, equipment or work systems can often make them much safer.
- Substitute with less hazardous materials, equipment or substances or eliminate the hazard.
- Effective ventilation—use adequate ventilation systems.

Other controls

- Administration—job rotation or timing jobs so less exposure to hazards.
- Training—hazards and safe working procedures.
- Maintenance and house-keeping procedures.
- Personal protective equipment—supply and maintain as well as train in correct use.

It should be emphasised that the hazard should be eliminated wherever possible before the need for personal protective equipment or signs are used. It is only when it is impractical to eliminate the hazard that these need be used.

It is the employer's responsibility to provide and maintain suitable personal protective equipment e.g. safety glasses, ear muffs, hard hats etc. as well as provide suitable training in their use.

HAZARD REPORT FORMS

All workplaces should have a hazard reporting system in place for workers to report potential hazards. This will bring problem areas to the attention of management as soon as the hazard has been identified.

Standardised hazard report forms should be readily available to the workforce. Workers should complete a form and give it to their immediate supervisor as soon as a potential hazard is identified. This will allow control measures to be put in place to remove the hazard at the earliest possible time.

WORKPLACE INSPECTIONS

These are regular inspections of the workplace to determine what hazards exist by observation. They are made by management and representatives of the workforce. Reports and recommendations from these inspections will allow for hazard control measures to be undertaken.

At workplaces with more than twenty employees, permanent safety committees are normally established, with representatives from management and the workforce. These safety committees are then responsible for carrying out the inspections.

ACCIDENT REPORTING

All workers at a place of work have a responsibility to report any illness, accident or near misses they are involved in or see. They should report these incidents to their immediate supervisor, site supervisor, a member of the workplace OHS committee, a union delegate or first aid officer. If there is any plant or equipment involved they must make the operator aware of the problem immediately.

Standards Australia's AS 1885.1–1990, Workplace Injury and Disease Recording, provides a national standard for describing and reporting occupational injuries and disease. From this national standard each state develops an accident reporting register and reporting procedures to suit its requirements under its own OHS legislation.

The state authorities may require serious work-related illnesses, injuries or dangerous occurrences to be reported on an accident report form. The accidents which must be reported by law are known as reportable accidents.

QUESTIONS

1. *What is architectural design?*
2. *What is the function of an architect in relation to the owner and builder?*
3. *What are the plans of a building? Write a brief account of why they are necessary and who requires them.*
4. *What information must be shown on the plans of a building?*
5. *What are specifications? What information do they give that is not shown on plans?*
6. *What authorities must be consulted regarding the proposed erection of a building?*
7. *What is the code on which New South Wales local council authorities base their building regulations?*
8. *Describe the physical factors to be considered when planning the building of a house.*
9. *Discuss the difference between 'stress' and 'strain'.*
10. *Show with a sketch how the forces of tension, shear and compression operate in a beam.*
11. *Briefly outline the responsibility of the builder in building operations.*
12. *Describe the preparation of a site for setting out.*

13. Describe the use and importance of a datum peg.

14. Describe the operation of setting out a site.

15. What is reinforced concrete? State its constituents and the position of the reinforcement.

16. Where is a damp-proof course placed in a wall and why is it so placed? Name a material commonly used for this course.

17. Describe the use of flashing for windows. Illustrate your answer with sketches.

18. Explain the difference between two types of brick bonds. Make a neat sketch of each.

19. Explain the meaning of the term 'stepped' as applied to foundations. What determines the amount of stepping for a brick wall?

20. What are the advantages of building on a concrete slab as compared with other types of foundations?

21. Describe three types of doors used in a house or garage.

22. Where are the following used: architrave, skirting, doorjamb, and cornice?

23. Name three types of roofs and make a sketch of each.

24. What is the difference between a wall and a partition?

25. Make a neat sketch to show one method of finishing the overhanging eave of a tiled roof.

26. Make neat sketches to show the structural members of a gable roof and a skillion roof.

27. Make a neat sketch through a section of a timber wall to show brick piers and all structural members to the ceiling joists.

28. Show by a sketch how the lengths and angles for a common rafter are obtained by using a right-angled triangle.

29. Show by line drawings three types of roof trusses.

30. What are the meanings of the following terms: head, sill, reveal, lintel, barge, fascia, overhang?

31. Sketch and name three different types of hinges. State the most common use of each and explain why each is particularly suitable for the purpose indicated.

32. Describe the fitting of butt hinges, stating how their position is marked on door and frame.

33. What are the main physical problems to be taken into account when designing a house on sloping land? With the aid of sketches, describe how they are treated.

34. What is the responsibility of the owner or builder regarding local government requirements as to building plans and specifications?

35. In preparing a plan what scales are used and why?

36. Describe how the various members of a timber-framed wall should be set out and the types of joints used.

37. What are the positions of the following members: top plate, stud, nogging, and brace?

38. Sketch a roof suitable for a garage, naming the members.

39. Name several types of roof coverings and how they are fixed.

40. Make a neat sketch showing one method by which the edge cut of a jack rafter can be determined.

41. Make a sectional sketch to show a door frame and necessary architraves.

42. Describe the position of the following door members: ledge, brace.
43. Briefly describe the preparation, priming and finishing for painted woodwork.
44. When should stopping with putty be done, and why?
45. Why is it important to punch all nails in woodwork to be painted? Name the type of nail used for this work.
46. What is the difference between a primer and an undercoat?
47. What does OHS stand for on a building site?
48. When was the national Occupational Health and Safety Act established?
49. What must employers provide for the health, safety and welfare of their employees at work?
50. What are the responsibilities of the employees on a building site?
51. Name four common hazards that influence health and safety in the building industry.
52. List three physical hazards on a building site.
53. Describe what a workplace hazard is.
54. What are engineering controls of workplace hazards?
55. When is there a need for personal protective equipment or signs on a building site?
56. Whose responsibility is it to report an accident or illness on a work site?
57. What is a reportable accident on a building site?

SURFING THE NET

1. **Timber, building in Australia**

   http://oak.arch.utas.edu.au

2. **Timber for building (Timber Development Association)**

   http://www.timber.net.au

3. **Timber building in Australia**

   http://www.arch.utas.edu.au

4. **Health and safety issues**

   http://www.worksafe.gov.au

5. **Workcover NSW**

   http://www.workcover.nsw.gov.au

6. **National Occupational Health and Safety Commission**

   http://www.nohsc.gov.au/

Chapter 4

Spray Finishing

Many of the quick-drying finishing materials for woodwork are best applied by spraying. Aerosol finishes certainly warrant consideration for small jobs as they will give a professional finish if used correctly. Simply follow the directions printed on the can.

Spray-gun equipment can be used to apply stains, toners and fillers as well as clear and opaque finishes.

SPRAYING EQUIPMENT

Sophisticated computer-controlled equipment is used in industry for curtain coating and electrostatic spraying (Figs 4.1, 4.2). However, this chapter is designed to cover conventional low- and high-pressure spraying equipment and its use.

COMPRESSORS

The compressor takes air from the atmosphere, compresses it, and supplies the air pressure needed to operate a spray-gun.

Low-pressure air compressors are usually small portable units. A diaphragm or small piston-type pump delivers compressed air directly to the air line without the use of a tank (Fig. 4.3). This type of compressor is designed to be used with a pressure-feed

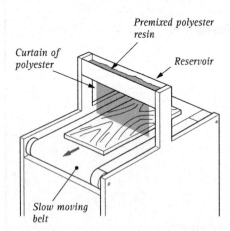

Figure 4.1 Curtain-coating equipment is used in industry to lay a heavy coat of polyester plastic finish on flat-faced furniture

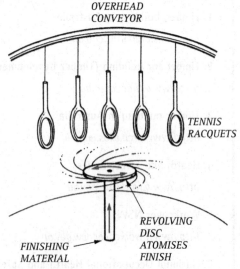

Figure 4.2 Electrostatic spraying: finishing material is atomised by high-speed revolving disc and then deposited electrostatically onto surface being finished

spray-gun. It can be used to spray most types of materials, especially those with a higher viscosity as the material is forced from the cup or container by direct air pressure.

Figure 4.3 Diaphragm-type portable air compressor supplies sufficient compressed air for low-pressure or pressure-feed gun

Compressors for high-pressure or siphon-type spray-guns need to supply a continuous high volume of air. Most larger-capacity air compressors have multipistons and a large reserve tank as they need to be able to supply at least 0.20 to 0.25 m^3/min at 200 to 280 kPa (Fig. 4.4).

Figure 4.4 Multi-cylinder compressor with storage capacity of at least 0.25 m^3 is desirable for spraying thin materials such as lacquer

Regardless of the type of compressor, clean dry air is needed. Water in the air supply can cause the lacquer to bloom (a whitish blush) while oil blown past the pistons can cause spotting. Air from the compressor should go into an air condenser (Fig. 4.5) to remove moisture, oil or dust. The condenser and the compressor tank should be drained frequently.

Figure 4.5 Air condensor traps water vapour and oil before they enter air line

The condenser is usually equipped with two gauges. One shows the pressure in the compressor tank—this should always be higher than that required for spraying in order that the actual spraying pressure is kept even. The second gauge is regulated to the pressure required at the spray-gun, which will depend on the material being sprayed and the type of gun being used.

A high-pressure hose is connected from the condenser to the gun and the length and inside diameter of the hose need to be considered in determining the required pressure at the gun. A hose, 10 m long and 6 mm in diameter, will lose approximately 30% of its pressure through friction.

Large stationary compressors, connected to a system whereby the air is piped around the workshop through galvanised piping, require built-in traps to remove oil and water—one trap at the outlet of the compressor and one at each outlet in the workshop. The compressor should also be downhill from the pipes so that the condensed water and oil will run back to the tank where they can be drained.

SPRAY-GUNS

The low-pressure or pressure-feed spray-gun allows air to pass through the gun at all times. It uses a high volume of air at low pressure.

The material being sprayed is forced from the cup or container by direct air pressure. This type of gun is used for spraying thicker materials (of higher viscosity) such as polyurethane varnishes and enamels. The trigger controls finish flow.

The principal advantage of the low-pressure gun is that less thinning of material is required. This in turn reduces loss of material

through overspray and reduces sagging or runs.

The high-pressure or siphon-feed gun is probably the most widely used gun in industry (Fig. 4.6). It uses a smaller volume of air at a higher pressure than the pressure-feed gun. The material is sucked from the cup with a siphon action and then atomised by the airstream outside the nozzle (Fig. 4.7). The air cap can be rotated to alter the direction of the fan pattern. There is also an adjusting screw to alter the width of the spray fan. Most manufacturers supply a range of needle, nozzle and air cap combinations to suit the type of material being sprayed.

SAFETY

The fumes from lacquers and epoxy finishes can be extremely hazardous as they are often both toxic and highly inflammable. The vapour of most solvents should never be inhaled for any length of time because they can injure the respiratory system.

1. *Never spray without using a proper respirator (Fig. 4.9).*

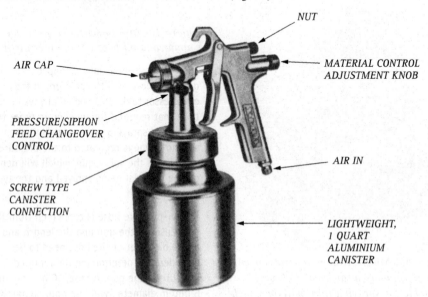

NUT

AIR CAP

MATERIAL CONTROL
ADJUSTMENT KNOB

PRESSURE/SIPHON
FEED CHANGEOVER
CONTROL

AIR IN

SCREW TYPE
CANISTER
CONNECTION

LIGHTWEIGHT,
1 QUART
ALUMINIUM
CANISTER

Figure 4.6 High-pressure or siphon-feed spray-gun

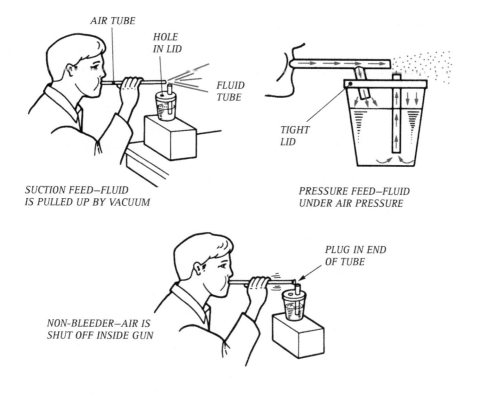

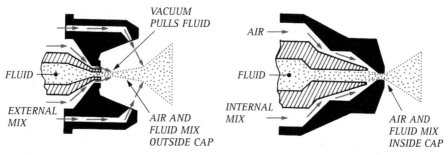

Figure 4.7 Principles of feeding and mixing finishes in spraying

2. Do not spray in confined spaces unless suitable extraction fans are used. The finishing area should have proper lighting, ventilation and exhaust systems.

3. Vapour- and arc-proof switches, lights and motors should be installed in finishing areas.

4. Containers of solvents and finishing materials should be securely sealed in fireproof metal containers and stored in a fireproof cabinet or room.

5. Oily paint or solvent rags are highly inflammable and should be stored in a fireproof container until they can be disposed of.

6. Proper firefighting equipment should be readily available including a fireproof blanket

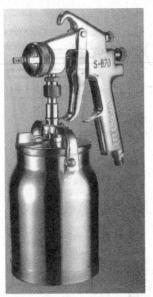

Figure 4.8 A slight vacuum is created at fluid tip as compressed air flows past, siphoning liquid from cup

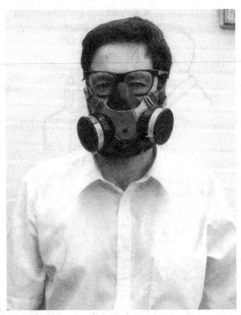

Figure 4.9 Proper respirators give protection against toxic fumes

to cover the fire at its source and water, foam, dry chemical or soda-acid extinguishers. Do not use carbon tetrachloride (CTC) extinguishers as they can produce a deadly gas with certain plastic fires. The soda-acid extinguisher is suitable for electrical fires. Never use water on electrical fires. Check with your local fire department for advice on the selection of extinguishers for your particular requirements.

SPRAYING PROCEDURE

Small jobs can be safely sprayed outdoors, if there is no wind, or in large well-ventilated indoor areas. A spray booth, however, is the best alternative where consistent quality and safety are required (Fig. 4.10).

Before spray finishing can begin, all pieces of furniture should be completely sanded. Allowance should be made for the thickness of material to be applied so that drawers and

doors will still fit after finishing. Where open-pored timbers require filling and paste-grained fillers are to be used, the task is best done by hand with the furniture pieces disassembled. The filling task can also be achieved by spraying on sanding sealers.

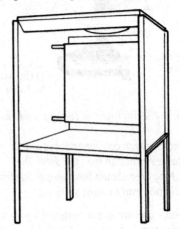

Figure 4.10 Hoods help to concentrate overspray which can then be exhausted from spraying area

The actual spray application of finishing coats can be done with the furniture basically assembled. Bits and pieces, such as drawers, loose shelves and handles, are best supported on a fixture such as a rotating table (Fig. 4.11). Small objects, such as handles, should be fixed to a board so they do not blow away.

Always pour finishes into the cup or container through a straining fabric, such as an old nylon stocking, to remove particles that could clog the gun. Thin the material being sprayed to the manufacturer's specifications and test the spray pattern on a piece of waste material. The finish should flow out smoothly and the application should be as thick as possible without sagging.

Figure 4.11 A rotating table allows the operator to make best use of available light; spacer blocks allow objects to be uniformly coated

Start the compressor and set the regulator to a higher pressure than required at the gun. A certain amount of air pressure is lost through friction in the air hose. The actual drop in pressure depends upon the internal diameter and length of hose. A general rule of thumb is to expect a pressure drop of 25% to 30%.

Regulate the fluid-adjusting screw on siphon-feed guns to control the amount of finishing material required. The trigger controls the amount of material on low-pressure guns (there is an adjusting screw behind the needle).

Adjust the air cap to produce the required external mix spray or fan pattern. This pattern can be adjusted to give a round spot or a wide elliptical spread (Fig. 4.12). Test the resulting spray pattern and viscosity of material on a sheet of paper or piece of scrap. Adjust if necessary.

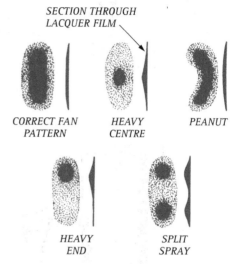

Figure 4.12 Correct and faulty spray patterns

Start to spray in exactly the same place every time. Always lap the same distance, move the gun at the same rate of speed and keep the gun at 90° to and at the same distance from the surface to be sprayed at all times. If these points are followed you will find that you will be making the same number of passes each time. Alter your speed or distance to get a perfectly painted surface. Once you have achieved the proper number of strokes at the proper speed, you have determined the proper spray application motion for the job at hand.

The spray-gun stroke is made by moving the gun parallel to the work and at a right angle to the surface. The distance from the gun to work should be from 150–200 mm for most paints (Fig. 4.13). Work with straight, uniform strokes, moving the gun across the surface in

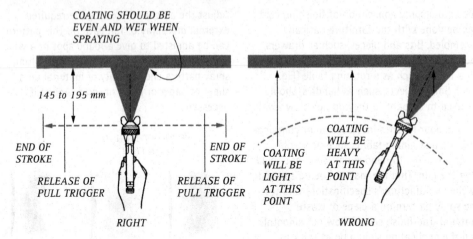

Figure 4.13 Keep spray parallel to surface to maintain even coating

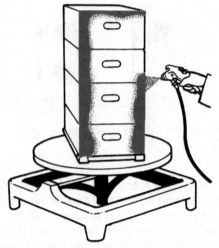

Figure 4.14 Corners of objects should be sprayed first

such a way that the spray pattern overlaps the previous stroke by 50%. Sags or runs may occur if the gun is moved too slowly or held too close to the work, or if the material has been thinned too much. A dry, sandy finish or orange peel effect is invariably caused by insufficient thinner or holding the gun too far away from the job; the material atomises and too much finish is wasted in the form of mist. A steady, deliberate pass that leaves a full, wet coat gives the best result.

The corners of objects should be sprayed first. When spraying flat surfaces each coat should be applied at right angles to the preceding one to ensure a uniform streak-free finish. Round objects, such as table legs, are best sprayed using a round rather than fan-type pattern.

If the trigger is pulled at the start of each pass and released at the finish, over-spraying is reduced and the operator has more control.

CLEANING THE SPRAY-GUN

Careful handling, cleaning and regular maintenance of the spray-gun are important aspects of successful spray finishing. The fluid needle packing, the trigger pivot bearing and the air-valve stem should have a regular spot of oil to keep them working freely.

Clean the gun as follows:

1. *Loosen the air cap slightly and remove the cup.*

2. *Pull the trigger to release any finish in the gun and let it flow back into the cup (Fig. 4.15).*

3. *Empty the finishing material out of the cup and wash the remaining finish out. Wipe the*

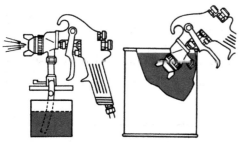

Figure 4.15 Spray thinner through spray-gun passageways to clean out all finish material

cup clean, using clean thinner and a lint-free cloth.

4. Spray the thinner through the gun to clear all passageways.

5. Remove the cap and tip and wash them thoroughly in fresh thinner. Use a small paintbrush to help clean each part. (See Fig. 4.16.)

6. Assemble the gun ready for use and store away in a dust-free area.

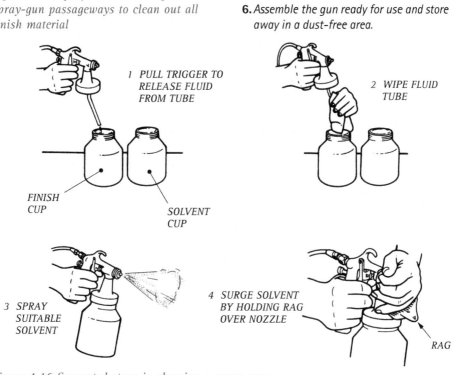

1 PULL TRIGGER TO RELEASE FLUID FROM TUBE

2 WIPE FLUID TUBE

FINISH CUP

SOLVENT CUP

3 SPRAY SUITABLE SOLVENT

4 SURGE SOLVENT BY HOLDING RAG OVER NOZZLE

RAG

Figure 4.16 Suggested steps in cleaning a spray-gun

QUESTIONS

1. Name the finishing materials that are best applied with: (a) a low-pressure or pressure-feed spray-gun; (b) a high-pressure or siphon-feed spray-gun.
2. Explain how a siphon-feed gun mixes and atomises the finishing material.
3. Why is it important to keep the vent hole clear in a siphon-feed spray-gun?
4. Explain the importance of storing oily or paint-saturated rags in a covered metal container.
5. What controls the width of fan or pattern of spray on spray-guns?

6. Explain the cause of and remedy for orange peel and sandy finish when spraying.
7. Moisture in the air line can cause blush or bloom. How can it be avoided?
8. Why is it important to hold the spray-gun perpendicular to the surface being sprayed?
9. The spray gun is a precision tool. List the steps necessary to clean and maintain it in first-class working order.

Chapter 5
Advances in Woodturning

Unlike other woodworking machines, a lathe is rarely used to merely process a workpiece from one stage of production to the next.

Complete objects, from a rough blank of timber through to a finished article, can be created on the one machine.

Woodturning is far more than a simple machining process—at its best it becomes an art form. Successful turning requires not only the mastery of very special techniques, but also an appreciation of what constitutes a pleasing shape with flowing lines.

The interest in woodturning and the proliferation of clubs around the world during the later part of the nineteenth century has no doubt led to many significant advances in the design of lathes, tools and accessories. This in turn has led to the recognition of woodturning as an artform.

DEVELOPMENTS IN LATHE DESIGN

SPEED CONTROL

Most lathes use a belt drive to transmit power from a 375 to a 750 W electric motor, using step cone pulleys to give speeds from 400 to 2800 rpm at the headstock spindle. In the past some lathes did provide a mechanical variable speed but recent developments in electronic speed controls, and reduction in the cost of producing them, give the woodturner better control of balance and spindle speed.

Recent developments in electronic speed control units provide high torque at low and high speed settings.

A lathe is specified according to the maximum length of workpiece, or distance between centres, and also according to its 'swing'—the maximum diameter of workpiece that can be turned over the lathe bed.

Lathes with a swivel head are a fairly recent development and they give the woodturner better access to the job as well as the capacity

Figure 5.1 A three phase motor is used for the speed control unit which converts the power back to single phase making it ideal for home workshop use

to turn larger diameter work than the lathe can normally produce. Generally though this requires the use of an outrigger or special tool rest holders for off end turning.

The lathe shown in Fig. 5.2 has both a swivel and a sliding head which increases the diameter turning capacity enormously, without the need for an outrigger. For example even though the spindle height is only 180 mm above the bed, platters and bowls up to 800 mm can be accommodated.

Figure 5.2 This lathe comes with two 600 mm long beds, which bolt together, giving a capacity of one metre between centres. Extra beds are available to increase the capacity of the lathe

Figure 5.3 The lathe set up in its shortbed mode. Five pulley speeds are provided, with double row angular contact bearings. Note the see-through headstock cover for speed identification

Figure 5.4 Set-up showing tool rest reversed for turning the underside of large bowls or platters

Figure 5.5 Turning the inside of bowls or platters. Slide the headstock along to the right and rotate it clockwise to a convenient position for turning. Position the toolrest holder and toolrest on the left hand end of the lathe and proceed to turn

WOODTURNING CHUCKS

When turning items such as egg cups, bowls or vases that have to be hollowed out, you need to remove the tailstock in order to turn the end grain. Consequently, the item or job must be held securely at one end by one of several special chucks mounted on the headstock spindle. Perhaps the simplest one of these is the screw chuck which is driven into a pre-drilled hole in the workpiece.

A cup chuck has a hollow recess with a very fine taper into which a matching conical spiggot has to fit. Some cup chucks have screw holes to help overcome the problem of the workpiece tending to work loose.

Figure 5.7 The screw cup chuck in many respects is much like a metal threading die except that a special thread form, that will cut a thread in wood, is used to thread a spiggot that has been turned on the end of the workpiece. This is a very accurate chuck providing it is screwed up firmly against the shoulder. This allows the job, in the case of a goblet for instance, to be removed and replaced very accurately during the finishing operation

Figure 5.6 The spiggot chuck shown is actually a large screw chuck designed to hold burls and heavy pieces on the lathe

SCROLL CHUCKS

Scroll chucks in many respects were derived in principal from self centring metalworking chucks in that the jaws are actuated by an internal scroll. This allows the jaws to be moved in and out so that the job may be held internally or externally. These ingeniously engineered devices incorporate not only a woodscrew, pin and cup chuck facilities but also contracting collets that can grip a cylindrical job or expand to fit a recess turned in the base of a bowl or a similar object.

Figure 5.8 A range of scroll chucks, both lever and key actuated, manufactured in Australia by Vicmark

LEADY CHUCK

This chuck was originally designed for bowl and box turning using the same principal as the friction or jam chuck.

This chuck has jaws contained in a circular groove in the backing plate. A centreless cutter, which cuts a cylindrical recess, allows the jaws to grip in a 3 to 6 mm deep recess.

(a)

(b)

Figure 5.9 Vicmark also produce a range of special jaws for their scroll chucks that can be readily attached for different situations (a) bowl jaws; (b) shark jaws

Figure 5.10 The cone-shaped centre expands the jaws when the chuck is turned in a clockwise direction. It is self-tightening when the job is loaded onto the chuck and rotated. The job is simply removed by rotating it in an anticlockwise direction whilst the spindle is stationary

Figure 5.11 Sixty millimetre diameter centreless cutter and chuck designed to a size that functions well on both small and large turnings

ECCENTRIC CHUCK

The 'Eccentric' chuck is basically a screw cup chuck that has the facility to swivel or offset the workpiece at a variety of preset angles to produce unusual eccentric turnings.

Figure 5.12 The chuck comes with a gauge that gives the length and diameter of the spigot that has to be turned on the workpiece before it is screwed into the threaded cup; the Allen key is also supplied to lock the chuck to a particular marked setting

Each setting is positively located with two fixed pins and a 10 mm bolt so that designs created can be accurately repeated. For example in the turning of cabriole legs for furniture.

There are a few basic rules to be kept in mind when using the eccentric chuck. When the job is offset there will be a certain amount of imbalance and this can be controlled by using the tailstock for support wherever possible and by selecting a suitable speed. Generally one speed slower than you would use normally on balanced work is satisfactory.

Figure 5.13 This chuck opens a whole new area for artistic woodturning

Figure 5.14 This chuck opens a whole new world for the creative woodturner, but also functions as a standard screw cup chuck

Successful offset turning involves the recognition by the turner that when the job is rotating it appears to show a solid core and a see-through shadow. This acts as visual guide for the turner but keep in mind that if you want to cut the wood the way it wants to cut then you must control the cut by keeping the bevel on the cutting tool rubbing. Don't forget the tool is only touching the job some of the time during each revolution when turning the shadow.

SPECIAL TOOLS AND ACCESSORIES

CUTTING MULTIPLE BOWLS FROM A BLANK

A number of bowls can be cut from a blank using a straight parting tool but a large amount of waste is still wasted as each cut only produces a conical blank. During the late 1800s and early 1900s the common method of parting bowls was to use a curved parting tool supported by an extra handle. See Fig. 5.15.

During the last two or three decades a number of commercially produced multiple bowl turning accessories, such as the Stewart system have been produced. The Stewart system included a number of straight and curved parting tools that could be attached to a handle that cradled your arm for support.

Figure 5.15 The old fashioned curved parting tool was exciting to use

The 'bowl miser'

This accessory designed by Bruce Leadbeatter was so named because of its ability to save the cost of timber. There are now a number of other similar devices on the market to serve this purpose.

The bowl miser uses a single curved parting tool with a 100 mm radius that enables the operator to part a number of different curved bowls by simply changing the angle of entry.

The guide for the cutting tool fits into the toolrest holder and is designed to control the twisting forces and the height of the cutting edge during the parting operation.

A relatively slow cutting speed gives more control and best results.

Remember when cutting two or three bowls from a blank to make provision for re-chucking by cutting a suitable recess.

Figure 5.18 A nest of bowls cut from a single blank

Figure 5.16 The aiming device shown helps the turner to accurately control the shape and size of each bowl being parted off

Figure 5.17 A small pumping action on the handle helps to clear the shavings from the cut. Allow for shrinkage when using green timber

VASE HOLLOWING TOOL

The normal procedure for turning vases involves chucking the workpiece securely, turning the outside shape, and hollowing the inside to a uniform thickness. Most vases are turned from timber with the grain parallel to the lathe spindle. Small vases may be turned

being held in a cup chuck, screw chuck or scroll chuck, but large vases require the use of a scroll chuck with special long jaws.

Simply screwing the timber to a relatively small face plate with fairly long screws will control the cantilever forces involved, when turning large vases.

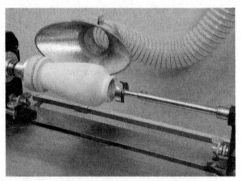

Figure 5.19 Using a forstner bit and an extension bar to drill out the bulk of the waste from inside the vase

Whilst a simple scraping tool can be used instead of drilling, drilling is a more efficient method for hollowing. Use a small bit first then open the hole further with a larger bit—this is necessary because the cutting action is against the grain.

The cutter in this tool is angled so that it will cut all of the internal surface of the vase. The gauge support bar is designed to rest on the toolrest giving two-handed control. It is just a

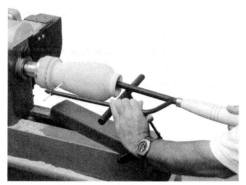

Figure 5.20 The vase hollowing tool has a curved end to which a variety of tips can be attached; the flexible gauge helps to control wall thickness

Figure 5.21 First cut is made from the right so the second cut pushes the ring off the disc automatically so that it rests out of the way on the ringcutter

matter of cutting the inside of the vase in a direction so that the fibres tend to lie down.

Shavings are best removed with a vacuum cleaner.

THE RINGCUTTER

This device is an attachment for woodturning lathes designed to cut rings from wood, plywood or MDF (medium density fibreboard) to be glued together to form the basis for turning bowls, vases, canisters, jewel boxes or even picture frames and bangles.

Basically, the ringcutter is machined to fit into the tool rest holder, and two fine high speed parting tool blades cut alternately into each side of a disc to produce a ring. The cutters are set at centre height and can be set at any required angle.

The device has a number of advantages over using solid wood: it makes use of scrap offcuts of wood; minimises waste of timber; size of planned turning is limited only by the size of the lathe; eliminates the hollowing operation when turning vases, bowls etc.

Planning your turning

The ringcutter was designed to cut 20 mm or less thickness of material. A 300 mm × 19 mm

disc with rings cut at an angle of 55° every 12 mm, will provide a series of rings which when reversed and glued together will produce a fruit bowl with almost no waste. See Fig. 5.23.

A more rounded shape of bowl would require a half section drawing and the use of two or more discs, by cutting the first ring from disc 1, the second from disc 2 and the third from disc 1 and so on . . .

For stability and appearance each ring is glued with the grain running at right angles to the last one.

Large vases, bowls or turnings are best built-up in stages, turned, and if necessary, finished on the inside before being finally assembled

Figure 5.22 A disc sander being used to produce a flat face on each side of the ring

Figure 5.23 Rub glue each successive ring, working from largest to smallest. Use weights or a simple press as shown.

and finished between centres. In the case of a vase a chuck would be fixed to the base and a tapered plug in the top supported by the tailstock. See Fig. 5.2.

Figure 5.24 Samples of articles turned from offcuts of red cedar, radiata pine, and sapele mahogany using the ringcutter

Large items require a certain amount of planning. A half sectional drawing is shown in Fig. 5.25, with rings numbered in sequence and lines showing the width of each ring.

The largest rings are best cut first so that other rings can be cut from the remaining centre piece.

Large vases are constructed by simply turning a number of bowls and gluing them

Figure 5.25 When using the ringcutter the size of the object being turned is limited only by the capacity of the lathe being used

together. Each bowl except the bottom one requires a sacrificial base for chucking. A one or two millimetre rebate turned between each section helps for concentric assembly, but is not absolutely necessary if care is taken.

BALL TURNING ATTACHMENT

The ball turning attachment fits into the tool rest holder and uses a specially sharpened scraping tool to cut the ball. Spheres are cut by securing the fulcrum directly under the centre axis of the lathe.

Elliptical shapes are turned by offsetting the fulcrum, and the inside of bowls trimmed to a selected radius by reversing the tool in the attachment.

Make sure the attachment is exactly on centre by rotating the tool from back to front of the ball. The tool needs to be sharpened along the left hand side to remove the waste to the left of the ball.

DRILL GUIDE ATTACHMENT

The drill guide fits into the tool rest holder and may include a number of bushes to suit different size drills. It can be locked at any desired angle but can only be used on lathes

Figure 5.26 Using a high speed, sharp scraper with a burr gives a nice clean cut. Cut with the grain where possible.

Figure 5.27 Using a drilling jig to drill holes at a constant angle during the turning of a space age pen and pencil holder

that have spindle indexing. It is a handy device for drilling accurately placed holes in turned work. For example drilling a hole through the head of a round mallet—drill half way from each side to prevent tear-out. Or drilling holes for three legs on a table lamp.

GOBLET RECESS CUTTER

Special cutters can be made, some are commercially available, to cut a variety of hollows for goblets, egg cups and small boxes. A sharp cutter at low speed will produce a smooth, uniform hollow.

Figure 5.28 Using a 50 mm cutter to cut a uniform shaped recess in a goblet. The recess is cut into the timber before the outside shape is turned.

QUESTIONS

1. Describe the advantage of having a variable speed facility on a woodturning lathe.
2. List the advantages of using a lathe that has both a swivel and a sliding headstock.
3. Explain the different use between the screw cup chuck and the screw or spigot chuck.
4. Describe the procedure to be followed to produce the recess for a scroll chuck or the expanding chuck.
5. What procedures should be followed, when using the eccentric chuck, to control the inherent imbalance?
6. The flexible finger on the vase hollowing tool serves a special purpose. What is it?
7. When using the ringcutter, why is it preferable during assembly to glue each ring with the grain at 90° to the last one?
8. List the advantages of using the ringcutter instead of using solid wood.

SURFING THE NET

Sydney Woodturners Guild

http://www.sydneywoodturners.com.au/

Chapter 6
Carving and Wood Sculpture

Basic to success in woodcarving work are a knowledge of the use and sharpening of edge tools, an understanding of the grain characteristics of various kinds of wood and a certain flair for creative design. Study of the chapters on care and maintenance of tools and preparation of timber in Woodworking, Part One will help in developing the first two, but practical experience in carving is the greatest teacher of all. Practise the basic cuts using a variety of carving tools on different kinds of timber and study the results. See how the wood behaves when cut in certain ways. This will eventually build up the degree of insight that will make it possible to tell almost at a glance how any given piece of timber will 'work'.

WOODS FOR CARVING

Well-seasoned timber, free from cracks and knots, is essential. Wood that feels soft under the knife is not necessarily the best for carving, for it may be spongy and difficult to cut cleanly except with a very sharp, thin chisel. Look for a fine, close-grained timber which does not split easily—Queensland maple, which has the added advantage of pleasing colour, fills these requirements admirably, and so do some of the Pacific maples.

Wood varies so much in texture, even in the one piece, that no hard and fast rules can be laid down and each piece must be judged on its merits. Species that show considerable variations but may yield suitable pieces are Oregon pine, blackwood, Huon pine, and English and Pacific oak. Colonial beech cuts excellently but is uninteresting in appearance. Coachwood and colonial cedar are both fairly easy to cut, but the former splits readily and the latter, especially the pale-pink varieties, is often too soft.

TOOLS FOR CARVING

Chisels and gouges in various shapes and sizes (Figs 6.1, 6.2) are the chief tools. Blades are usually tanged into the handles and the total length is 175 to 250 mm— over-long tools are difficult to control.

Sharp knives are also useful, especially for chip carving. Other equipment includes a round mallet; G clamps and bench holdfasts (Fig. 6.3); small duster brush; tenon saw, bowsaw, coping saw, jigsaw, or bandsaw; spokeshave; smoothing plane; abrasive papers and cloths; files; oilstones, oilstone slips and a grinding wheel; leather strap for stropping edge tools; and, not least in importance, a firm bench.

SHARPENING WOODCARVING TOOLS

The feel and sound of woodcarving tools slicing through and curling effortlessly away has to be experienced to be appreciated. The importance of razor sharp tools cannot be overemphasised. Blunt tools are hard to use

Figure 6.1 A useful set of carving tools

Figure 6.3 Equipment for shaping wood: spokeshave, scraper and bench screw

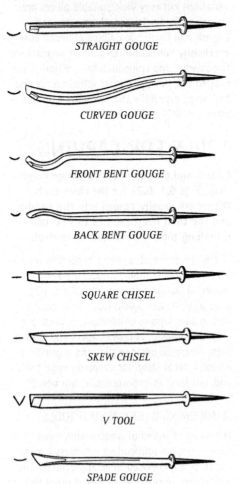

STRAIGHT GOUGE

CURVED GOUGE

FRONT BENT GOUGE

BACK BENT GOUGE

SQUARE CHISEL

SKEW CHISEL

V TOOL

SPADE GOUGE

Figure 6.2 Blade shapes of carving chisels and gouges

and produce poor results. Resharpen tools at the first sign of tearing or catching, give them a light hone or touch on the emery wheel.

Sharpening new tools

New carving tools are usually ground to shape, but not honed to a razor-sharp edge. Carving tools need to be honed on both sides of the cutting edge—be prepared to spend time until the wire edge is removed. The selection of fine grinding wheels and sharpening stones is important for this purpose.

Sharpening tools

The sharpening of woodcarving tools is different to that of chisels for cabinetmaking. Chisels for woodworking have a different grinding and sharpening angle whereas woodcarving and woodturning tools should be sharpened or honed at the same angle as the bevel.

The bevel on woodcarving tools is usually rounded so that the bevel controls the cut.

Using a fine water, oil or diamond stone place the bevel flat and draw the tool backwards, at the same time lowering the handle; then push the tool forward and lift the handle. Repeat this action until both sides are smooth and rounded and a fine burr or wire edge is produced. Remove the burr and polish the bevel to a razor-sharp edge by stropping on a leather strop or on a power buff.

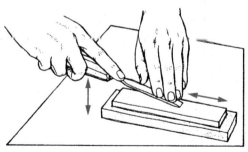

HONE THE BEVEL ON BOTH SIDES

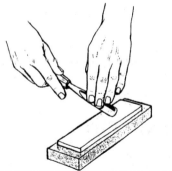

ROLL THE GOUGE FROM EDGE TO EDGE

HONE THE POINT WITH A SLIPSTONE

HONE INSIDE BEVEL WITH SLIPSTONE

Figure 6.4

PREPARING A DESIGN

The design must be appropriate to the object being decorated and to the type of timber being worked. To originate a design calls for imagination and the ability to form a mental picture of the finished product. For chip and incised carving this is relatively easy, since only outlines are required, but for pierced work, relief, and wood sculpture it is necessary to consider the effects of light and shade. It may be helpful to model a wood sculpture first in clay or plasticine.

Copying an existing design is simple enough. Mark a grid pattern on both the original and the workpiece. Then transfer the points where the lines of the design fall in the squares on the original to the squares on the workpiece.

To reduce the size of a pattern, make smaller squares on the workpiece; to increase the size, make the squares larger. Patterns on wallpaper, lace, wrought iron, pottery, leatherwork and textiles can all be adapted for wood. Many traditional carving designs can be found on old furniture or in books and magazines on art and furniture decoration. Patterns include Greek key, egg and dart, guilloche (interlaced circles and scrolls), Romayne (a head in a medallion) as well as numerous combinations of shells, leaves, fruits and birds, to name only a few.

Instead of setting out the design directly on the wood, draw it on paper first and then transfer it with carbon paper. Make a pattern for repetition ornament and then outline it on the work with a pencil. If the design is symmetrical, fold the paper in half, draw half the design, cut to shape and then unfold the paper.

Beginners are wise if they avoid small, intricate designs in favour of large, bold ones, which demand less skill and do not show faults so readily.

BASIC CUTS

All types of carving use the same basic cuts:

1. the stop-cut;
2. the slicing-cut made to the stop-cut.

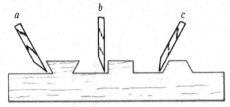

Figure 6.5 The stop-cut: (a) Undercutting is bad as edge tends to break; (b) A vertical cut is generally made first; (c) A slant-cut is desirable for narrow sections

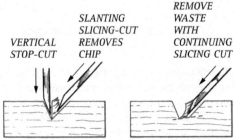

VERTICAL STOP-CUT   SLANTING SLICING-CUT REMOVES CHIP   REMOVE WASTE WITH CONTINUING SLICING CUT

Figure 6.6 The two basic cuts in all carving; in a long groove, several stop cuts may be needed to facilitate slicing out waste

1. VERTICAL STOP-CUT ACROSS THE GRAIN

2. SLICING-CUT IS MADE BY CHISELLING AT APPROX. 45°

3. ANOTHER SLICING-CUT IS THEN MADE ON OPPOSITE SIDE OF 'V' GROOVE

Figure 6.7 Stop-cuts for narrow relief

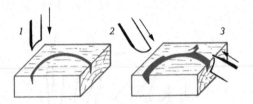

Figure 6.8 Carving a curve: (a) Make vertical stop-cut with firmer gouge; (b) Where curve lies across grain, use same tool at 45° for slicing-cut, (c) Drive tool along grain to make slicing-cut with grain

Figure 6.9 (a) Chiselling with grain-tool tends to rise, taking clean chip or shaving; (b) Chiselling against grain-tool tends to 'dive', and chip tears away

Figure 6.10 Slicing action to stop cut along grain

Stop-cuts are made square, or at a slight slope, to the face and should not be undercut (Figs 6.5, 6.6). In a long groove, several may be necessary to facilitate slicing out the waste (Figs 6.7, 6.8).

Relief carving requires stop-cuts, made on a slope so that the base is slightly larger than the outline, to delineate all parts that are to be left standing above the background. Removal of waste to form the background is done more quickly by cutting across the grain than by cutting with the grain (Figs 6.9, 6.10).

Study Figures 6.5 to 6.10, then practise making cuts in a piece of wood cramped to the bench (not held in the vice). The mallet is used with chisels and gouges for most work, but in some cases the cutting tool is worked with both hands, the right for pressure and the left as a guide.

TYPES OF CARVING

Woodcarving as decoration is of five types: chip, incised, pierced, relief, and in-the-round, the last often referred to as 'wood sculpture'.

CHIP CARVING

Chipped designs are always geometrical, featuring triangles, squares or circles (Fig. 6.11). Chisels are used in hard timber but a sharp, skew-blade knife is adequate for soft wood.

Figure 6.11 Electric carving tool makes chip carving easy

INCISED CARVING

Incising is particularly useful for carving letters. The design is incised into the surface with cuts of varying depth, width and shape, the original face forming the background. In another form, called 'chasing', the pattern covers the entire surface.

PIERCED CARVING

In pierced work, the background is cut out with a bowsaw, coping saw or jigsaw, leaving the design clearly outlined. In surface carving, the design is enhanced with incised or relief work. In level surface carving, the surface is left flat with no attempt at relief detail, though incising may be used to represent such features as leaf veins. Examples of pierced carving may be seen in frets and chair backs.

RELIEF CARVING

A relief carving is the most difficult to execute and demands a sound knowledge of timber properties. Considerable modelling skill is required to produce a high-relief carving, where the ornament, in the form of foliage, fruits, flowers and figures, is carved to actual shape. Low-relief carving is distinguished by a shallower background (Fig. 6.12).

Figure 6.12 Platypus carved in relief by Maricha Oxley

IN-THE-ROUND CARVING OR WOOD SCULPTURE

As the term 'in-the-round' implies, shapes are carved complete, not as applied ornament or part of a panel. Figures may be abstract, stylised, or natural and, like high-relief carving, the work demands a flair for modelling and for selecting timber that will lend itself to the required visual effect (Figs 6.13, 6.14, 6.15)

Fig 6.13 Carving in the round by Eric Satchell

Woodcarving goes beyond the utilitarian and offers the craftsman a means of self expression that raises it to the level of art.

The ability to produce a fine piece of work only comes with practice, though a natural eye for form is an undoubted advantage.

Designs for carving can be naturalistic or abstract, the concept being dictated by the carver and by the natural features of the wood.

Simple solid shapes are easier to carve successfully while you are developing your skills and techniques. Carvings that are pierced or deeply undercut leaving thin weak sections can be difficult for the beginner.

CARVING METHODS

SETTING OUT

An experienced woodcarver may simply select a block of wood and carve into it, the shape evolving as the wood is cut away following an imaginary outline. This can be a hit and miss process. A better approach is to work from full size drawings, showing front and side views. These views are then traced onto the block so that the bulk of waste can be gouged or sawn away.

Designs initially moulded in clay or plasticine can be of great help in the preparation of drawings.

Figure 6.14 Modelling in clay or plasticine gives a three-dimensional plan to work from

SHAPING

After the bulk of the waste has been removed the block needs to be fixed or held firmly in preparation for further carving. A solid bench and vice and a bench and holdfast are often all that is necessary.

A metalworking vice with soft jaws, especially one with a swivel base, is useful for the purpose.

There are a number of professionally manufactured holding devices such as the carver's bench screw and pivotting clamps (Fig. 6.17). A carver's vice is similar to an engineer's vice, but is made of wood and has cork- or leather-faced jaws to protect your carving.

Figure 6.15 Note the clay model in the background and the commercial holding device to hold the piece whilst carving

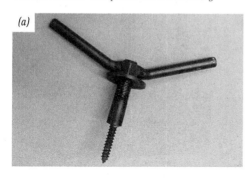

(a)

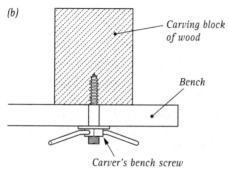

(b)

Carving block
of wood

Bench

Carver's bench screw

Figure 6.16 (a) Carver's bench screw
(b) example of use

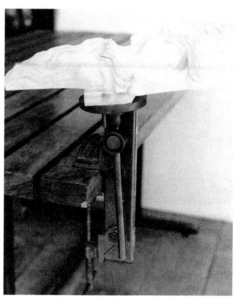

Figure 6.17 Record carving vice, pivots to a convenient angle.

HOLDING TOOLS

Carving is a craft that draws upon natural skills. A sense of proportion, co-ordination between hand and eye, the interpretation of materials and texture and a feel for a natural line all contribute to make a fine carving.

Holding and using the tools correctly is the first step.

Paring cuts are made with both hands on the tool.

Woodcarving differs from other methods of cutting wood in that slicing cuts are generally made across the grain. This is especially so when roughing out the carving.

As your carving progresses the wood may have to be cut at all angles. Look closely at the grain and make cuts in the direction which is least likely to tear the fibres.

Choose and set up a suitable block of wood with the grain lying in such a way that you can work mostly from top to bottom.

Figure 6.18 Electric carving tool with
interchangeable blades gives excellent
control

Figure 6.19 Emotional expressions by
Maricha Oxley

Since all surfaces except the bottom will show
in the finished piece, the problem of holding
the work during the carving arises.

There are several possible solutions:
- Glue or screw on a piece of wood or
  plywood as a temporary base for clamping.
- Leave extra wood at the bottom for vice or
  clamping, then later remove the excess.
- Use the woodcarver's clamp through the
  bench top.
- Use G clamps, bench holdfast, vice or bar
  clamps, but be careful not to bruise or mark
  the exposed surfaces.
- Sandbags may be arranged to cradle the
  work if all surfaces are exposed.

Figure 6.20 Delicate carvings demand a sense
of touch and a well thought out procedure

Figure 6.21 In-the-round carving showing
step-by-step procedure by Maricha Oxley

MARKING OUT AND CUTTING THE WOOD TO A ROUGH SHAPE

Detailed drawings may be of some use when making a woodcarving or a piece of sculpture, but a three-dimensional model that can be reworked or reshaped to provide an approximation of the final product will be of more help.

Figure 6.22 Clay modelling gives a more creative approach; sketch first, model then carve for best results

Materials for making three-dimensional models are clay, soap, plaster of Paris, plasticine, polystyrene foam, wax and even play dough.

Front, back, and side views may need to be drawn on the piece, but only in basic outline before the bulk of waste is removed.

The waste may be removed by bandsawing, chiselling, turning, scroll sawing, power cutters or even chainsawing depending on the size of the piece being worked.

FINISHES FOR CARVED WORK

Sometimes carved work is left with the tool marks showing to achieve a certain pattern effect. In this case great care must be taken with the carving if the finish is not to be mistaken simply for bad quality work, and no abrasives must be used to spoil the clean sharp lines. More often a smooth finish is desired and this is obtained by the careful use of abrasives and the application of finishes.

Figure 6.23 Remove the bulk of waste with gouge or bandsaw before carving

Surfaces may be textured by leaving tool marks, or smoothed by scraping or sanding. Use the effect that appeals to you.

Whilst a textured surface saves scraping and sanding, considerable thought and skill is required to create tool marks of suitable size, shape, direction to suggest the texture and detail desired.

Figure 6.24 Flowing lines in silver ash enhance the beauty of this piece

If a sanded surface is desired all tool marks should be removed before sanding commences.

A smooth finish should be applied to articles such as spoons, plates and bowls on food-contacting surfaces for ease of cleaning.

Outer or unused parts of course may be left with tool marks for textured effects.

Applied finishes are used for three reasons:

1. *to enrich the effect of the carving;*

2. *to preserve and bring out the natural beauty of the timber;*

3. *to make the article suitable for a particular use.*

As an example of the last, salad bowls and dishes for other foods are simply finished with a vegetable oil, such as olive, grapeseed or safflower. For other items, there are three different clear-finish treatments which give a subdued gloss:

1. *Apply several coats of linseed oil and follow with several applications of wax.*

2. *Brush on several thin coats of orange or bleached shellac. Note that this finish is not waterproof.*

3. *For a hard, durable finish, apply several coats of Swedish oil or one of the proprietary plastic coatings.*

Though not in line with the modern trend, colour may be introduced before clear-finishing by staining the timber, using oil stains rather than water stains, which tend to raise the grain. The background in relief carving is sometimes lightly stained to show up the design. Another method is to coat the piece with paint and wipe it off before it dries, leaving a residue of colour in the pores of the wood. Charring with a gas torch followed by a light rubbing down creates an interesting effect on free-form work (Fig. 6.26).

Full-gloss finishes are seldom used, since they produce highlights which spoil the general effect of carved work.

Figure 6.26 Corroboree—note the turned cone within—using triangle frames with flashing light within by Maricha Oxley

Figure 6.25 Clydesdales in golden cedar by Maricha Oxley

Figure 6.27 Beautiful leaf bowl in camphor laurel using incised carving by Ian Irwin

Fig. 6.28 Oregon pine carved pieces charred and wire brushed to give a textured look.

QUESTIONS

1. Name five types of woodcarving that are used as forms of decoration.
2. What timber characteristics should you look for when selecting wood for carving and sculpture?
3. List four Australian and two imported timbers suitable for carving or sculpture.
4. Describe some of the important points to be considered when creating your own design for a woodcarving.
5. Describe how you would transfer a traditional carving design onto a piece of furniture you are making.
6. Why are chip-carving designs almost always geometrical?
7. Why does the woodcarver invariably use a paring gouge across the grain when roughing out a hollow object such as a fruit bowl?
8. Sketch a stylised design for a piece of wood sculpture based on the form of a sleeping dog. Indicate the species of timber you would select for this piece, show the grain direction and nominate a suitable finish.
9. How do you remove the wire edge from a firmer gouge?
10. How does a high-relief carving differ from carving in-the-round?
11. What advantage does modelling in clay before carving give the design, rather than working from sketches?
12. How does the function of a bench holdfast differ from a bench screw for woodcarving?
13. List some of the things to be considered when selecting an appropriate finish for a carving or a piece of wood sculpture.
14. Describe what modifications would be necessary before a metalworking vice could be used to hold down your piece of sculpture.
15. What are the advantages of using colour or stain on carvings or sculpture before a clear finish is applied?

SURFING THE NET

Wood sculpture

http://www.woodart.com.au

Chapter 7
Historical Aspects

The full story of the development of the woodworker's craft can never be told, for though the earliest historians wrote copiously of wars, politics, religion and philosophies, they left no detailed records of the abilities or products of craftspeople. Yet the study is a fascinating one, perhaps for the very reason that it must be pieced together by inference and conjecture from scanty information, by reference to the tools devised to aid the craftsperson's handiwork, and how those tools were used in making the furniture for dwellings and the buildings to shelter the activities of daily life.

DEVELOPMENT OF HAND TOOLS

In museums all over the world are specimens of tools discovered in ancient tombs, excavations, marshes and middens. However, to fill the gaps in our knowledge of how and for what purposes these tools were used, we must study ancient murals and the decorations on tombs, coffins, pottery and sculpture.

There are literally thousands of examples of stone, copper and bronze tools still existing but very few of wood and iron, as these are subject to decay and rust. Iron and steel, however, are frequently mentioned in ancient literature. Of Solomon's Temple, the Bible says

that, '...there was neither hammer nor axe nor any tool of iron heard in the house, while it was in building'. Homer's account, written not later than 900 BC, of Ulysses driving the stake into the eye of Polyphemus states that 'the hissing was like steel quenched in water'. From this simile we can deduce that working in steel was a familiar process in Homer's day. While it is doubtful whether early peoples knew steel as we know it today, they were aware of the varying degrees of hardness of iron and tried to produce harder qualities. The method of smelting iron with charcoal certainly produced a form of steel, since a certain amount of carbon would be absorbed from the charcoal.

For convenience in placing events in point of time, history has been divided into ages or periods—the Stone Age, the Bronze Age, the Iron Age, the Greek, Roman, medieval, Renaissance and modern periods—but some of these overlap. Thus explorers like Columbus and Cook found people still living a Stone Age existence. Moreover the stages of development represented by these ages were not reached simultaneously by people in different geographic areas. For example, there is no evidence that the ancient Egyptians used planes, although they were well known to the Romans of the same era. Similarly the Romans and the Assyrians had compasses, but it would appear that they were not then used by the Greeks. The prevailing local level of culture, it seems, determined the kind of work the craftsperson did and the kind of tools used—there are indications that tools in common use in medieval English or French villages were neither better nor more varied than those employed by Roman craftspeople a thousand years before.

In brief, the development of tools of all kinds has been gradual, spasmodic and mainly dependent on the introduction of new materials and better methods of using them.

As with modern tool-making, the main problems for the ancient craftsperson were in producing good cutting edges and devising a means of fixing them in place so that the complete implement could be held and handled most comfortably and effectively. It is in the solution to these problems that we find the greatest advances in development and, as communications and transport improved, so the knowledge of improvements in tools was spread.

Much more rapid progress has been made since the middle of the nineteenth century, until which time tools were invariably handmade. Modern factories are capable, through their vast resources for research and production, of turning out tools of predetermined quality and excellence for almost any kind of work, while extensive advisory and publicity services quickly make users aware of the new products and their applications.

EDGE TOOLS

Ancient tools or implements of stone, copper or bronze are referred to as 'celts' (Fig. 7.1). Because they were not used for specialised purposes as we now use tools, it is impossible to state that any one tool, such as the plane or the chisel, was introduced at any particular time. Certainly the earliest cutting tools were

Figure 7.1 Stone Age flint chisel, 3000 BC

pieces of hard stone, bones or horn, depending on what was available in the area or could be obtained by barter from other people.

Authorities differ on whether very early humans simply chose their cutting tools from available pieces or sharpened them to produce a cutting edge, but it seems certain that at some stage, stone celts and also bone and horn were sharpened and shaped by chipping or grinding. The earliest copper and bronze celts were similar in shape to those of stone.

Axes and adzes

A most important advance was the invention of the haft, or handle, which gave the craftsperson more power and facility in the use of a tool and resulted in the forms of the axe, with the line of the cutting edge along the handle; and the adze, with the line at right angles to the handle (Fig. 7.2). Some of the earliest examples dating from 8000 BC have a cutting edge of reindeer antler with a pointed end like a pick.

Figure 7.2 Neolithic handled adze

The method of attaching the handle to stone and bronze heads was by tying with fibre or leather thongs, the latter often applied wet so that they tightened as they dried. The handle, a straight or bent piece of wood, was either notched underneath or split at the end to accept the head, but both types were weak at the point of impact as is evidenced by the number of specimens with broken handles. In some cases stone celts were bored and in others a groove was formed around the stone to help secure the binding. Later, copper and bronze celts were forged to shape with flanges on the sides bent round to hold and position the handle and pierced with holes to secure the binding. These flanged celts are known as 'palstaves'. Handles were not straight but selected with a crook at the end or cut at a fork, the short end being secured to the head.

The next development appears to have been the socketed head, the socket being formed opposite the cutting edge and still requiring a crook in the handle. The idea of making an eye for a straight handle does not seem to have occurred to early metalworkers, for although eyed stone celts have been found, none has been discovered in copper or bronze (Fig. 7.3).

Figure 7.3 Egyptian copper adze, seventeenth and eighteenth dynasties

Specimens of stone axes and adzes have been found in many parts of the world. Denmark and the ancient Swiss lake villages have yielded examples dating from 6000 BC, while others discovered in France and Wales were evidently in use 4000 years later. The Greeks were using stone axes in 900 BC, as were the American Indians when Europeans landed on their shores in the fifteenth century. Specimens of copper and bronze tools have been found in places as far apart as Peru, Egypt, Spain, England and Russia (Figs 7.4, 7.5, 7.6).

There is much evidence to show that early humans used fire to burn out waste wood before finishing the work with axe or adze. This is still a common method among some

Figure 7.4 Bronze axe

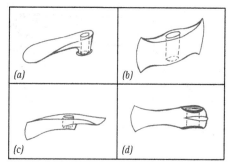

Figure 7.5 (a) Adze, Ur, 2700 BC; (b) Double-edged axe, Crete, 2000–1700 BC; (c) Axe-adze, 1700 BC; (d) Malvagni, 1000–800 BC

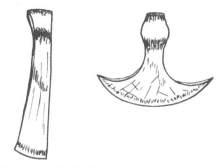

Figure 7.6 Typical medieval axes

indigenous peoples—Australian Aborigines living a traditional lifestyle hollow out logs for canoes by burning and they also harden wood with fire for making weapons.

Planes

It is likely that the plane developed from the adze. The earliest examples of planes discovered show an excellent standard of design, indicating that there must have been some transitional type of tool though no specimens have been found. The probable reason is that they were made of wood and iron and would soon be reduced to dust by decay and rust (Fig. 7.7).

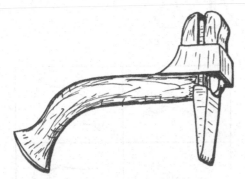

Figure 7.7 Iron-collared adze, showing collar and wedge

Specimens of well-designed and well-made planes, not much different from modern types, have been found in many countries. Strangely enough, they were evidently not possessed by the Egyptians, who are depicted in illustrations using chisels or adzes for shaping and stones as abrasives for finishing. Tradition has it that the Greeks invented the plane, Daedalus, a mythical craftsman of great skill, being suggested as the originator. Certainly the Greeks had a name for this tool, 'rhykane'—but so far, no Greek planes, in fact few Greek tools of any kind, have been found. It is also suggested that the Roman wedged adze (Fig. 7.8) used in Egypt could have been a type of plane, and another possibility is that planes developed from a kind of scratch stock through moulding planes to wide-ironed jack and smoothing planes.

The earliest types discovered are made of wood and sheathed with iron, with a single cutting iron held at an angle of 50° by a wedge positioned under a crossbar similar to the wedge in many present-day block planes (Fig. 7.9).

Excavations throughout Europe have yielded many examples of Roman planes, one at Cologne being completely of bronze with no wooden case, and iron specimens have been located in France, England, Germany, and

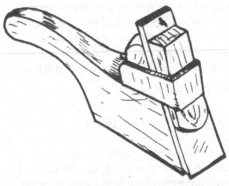

Figure 7.8 Modification of wedged adze

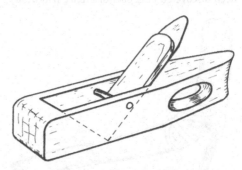

Figure 7.9 Roman plane from Pompeii

Scotland. The first Roman plane found dates from 50 BC and four planes found at Pompeii date from AD 79. Their dimensions are very similar to those of the Stanley No. 3 smoothing plane. A Roman plane found at Silchester in England is 336 mm long, 5.7 mm wide with a cutter 108 mm long set at 65° (Fig. 7.10). Another, a wooden jack plane, is longer than other specimens and is made of beechwood with a handle at each end (Fig. 7.11). A sandstone relief carving at Saarburgh, Germany, shows a similar tool in use. An interesting and unique find at the Roman fortress at Newstead, Scotland, yielded what appear to be the cutting and cap irons of a plane, although cap irons as such did not make a definite appearance until late in the eighteenth century.

From Roman times there are many gaps in the history of the plane and, until recent discoveries in Frisia, Holland, of planes used in the eighth and ninth centuries (the late

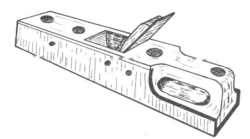

Figure 7.10 Reconstruction of Silchester plane

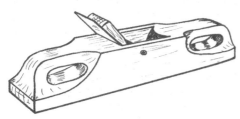

Figure 7.11 Roman wooden jack plane

Merovingian period), it seemed as though they had been forgotten and then rediscovered in the twelfth and thirteenth centuries. The Frisian planes were made of bone with no front handle, but a handle shaped in the stock at the back and the single cutting iron wedged at 40° to 45° under a crossbar. A similar type, of bone with a bronze sole, was found in Kent.

Few actual examples of planes have been found representing the period AD 800 to 1600, but pictorial references abound in carvings, stained-glass windows and illustrated manuscripts (Figs 7.12, 7.13). Notable among these are the thirteenth-century mosaic in the porch of St Marks, Venice; a painting by Ambrogio Lorenzetti and a fresco by Pietro Puccio at Pisa (AD 1390); inlaid panels by Agostine de Manchi in the choir stalls of the church at San Petrois, Bologna, Italy (1468–1477); a picture in the Bedford Book of Hours (fifteenth century manuscript); and a picture of a Paris joiner's shop by Bourdichons (1510). Durer's painting Melancholia (1514) shows a plane in detail. An important specimen, dating from 1570, in a collection at

Dresden has a small thumbscrew to hold the iron in place. The Novaya Zemlya plane (1596), from a Dutch expedition wrecked on the way to China, was found in 1871 together with a number of plane irons of various shapes.

Figure 7.12 Early German plane, probably sixteenth century

Figure 7.13 Sixteenth-century block plane

The centuries between 1600 and 1800 are sometimes called 'The Golden Age of Woodwork'. Rising standards of living, together with the increased travel and trade which brought new and exotic woods within the craftsperson's ken, called for a higher standard of skill and better tools in greater variety.

One of the first improvements was in the shape of handles, referred to by the term 'tote', or 'toat', whose derivation is obscure. Plane handles were formerly shaped out of the stock and let in horizontally or almost vertically or, as in some Tyrolean examples, took the form of leather straps nailed on behind the iron. Now, many planes were given an upright curved front handle similar to, but more elaborate than, that of the modern German jack plane. The name 'jack', incidentally, appears to have been used to denote a general-purpose or frequently-used plane. 'Jack knife', and 'lifting jack' are other examples of the application of this word to common or familiar things.

Planes of this period show much individuality in detail and decoration because most craftspeople made their own tools. Many were elaborately carved and have the date and owner's name or initials prominently shown.

Different types of planes for different purposes began to make their appearance in the seventeenth century. Joseph Moxon wrote in 1683 of the 'fore plaine which is used before the smooth plaine or joynter'. The modern foreplane is a shorter trying plane. Sticking the large-section mouldings then in fashion required a great amount of effort and for this work the craftsperson had an assistant, probably the apprentice, who pulled the plane by a rope attached to a hole in the front end or by means of additional handles (Fig. 7.14). Many specimens of such 'drag' planes may be seen in exhibitions, but with the introduction of the spindle moulding machine they were no longer necessary. Specialist craftspeople, such as musical instrument makers, coachbuilders, coopers and shipbuilders, developed planes for their own needs.

Wooden planes of the early nineteenth century, from which the front handle had been eliminated, closely resemble those of today. Only trying planes seem to have had cap irons, and since Moxon's time smoothing planes had been coffin shaped, of a type called in Europe the 'English pattern'.

While no accurate date can be set for the introduction of the cap iron, it appears that by about 1700 some planes had cap, or 'curling' irons. They were not fastened to the cutting

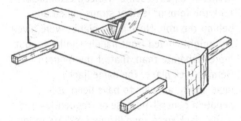

Figure 7.14 French 'galère', or 'drag',
plane; Demi Varlope, 1848

iron, but simply held in place by the wedge-irons and wedges fitting into a groove. From written records and references in the late eighteenth and early nineteenth centuries, it appears that the cap iron originated in England; the double iron is also mentioned in a French book of the period.

In the mid-1800s, many other important improvements began to take shape. Some idea of these can be gleaned from the records of patented ideas, though the picture is incomplete. Some of the ideas patented were not generally adopted and other very significant improvements were not patented; for example, it is difficult to determine when the now-standard practice of providing cutting irons with steel facings was introduced and who was the originator.

Many innovations were made in methods of holdings and adjusting the irons. In 1827, H. Knowles patented a cast-iron plane which had a channelled stock but no mechanical adjustment for setting, a single iron and wooden wedge being used. In 1843, W. Foster patented an improvement which regulated the thickness of the shaving—the cutting iron was not moved but the sole at the mouth was adjusted by a vertical screw. This idea, although sometimes followed today in conventional planes, was not popular. In 1844, T. Stanford patented a screw adjustment for moving the iron up or down, but not laterally. Further improvements were made by T. W. Worrell and later by Leonard Bailey.

Bailey's first patent in 1858 was for a friction-plate vertical adjustment which was ineffective, but the same patent also covered a cammed-lever cap. Early Bailey planes had a wooden stock, but Bailey is credited with many of the improvements which are now accepted features of metal planes, including bolting the handles to the bed; the curl in the end of the cap iron; a thin cutting iron of parallel thickness; and the adjustable frog. Bailey was a manufacturer of planes, but in

1869 he sold his business and most of his patents to the Stanley Rule & Level Co., later becoming head of its plane department.

Other inventors who contributed to the Bailey plane were G. H. Warren and J. A. Trant. Apart from inventing a rather expensive lateral adjustment, which was soon discarded, Warren devised the general arrangement of the plane. Trant invented the lever-type lateral adjustment and also redesigned the whole plane, giving it better appearance and balance and a more convenient arrangement of parts.

An idea to simplify honing, patented by the Record Company, is the Stay-set cap iron, the bottom portion being removable without unscrewing the iron. It is claimed to be firmer than one-piece cap irons and to reduce chatter.

In England and Europe, different ideas and innovations were being tried out at about the same time as American ideas. William Marples of Sheffield sought to improve on the American pattern and made their planes heavier, reducing the tendency to chatter.

One noticeable change in overall design is the reduction in the number of sizes—since 1870, the Stanley range has been reduced from fourteen to seven different lengths. Single special-purpose planes are now used mainly for rebating, ploughing and simple moulds, while modern electric routers with a variety of cutters are employed to produce other shapes (Fig. 7.15).

The development of special-purpose planes has a shorter history than that of bench planes, although the Romans are credited with using rebate planes. During the Renaissance period, when a wide variety of mouldings, intricate panels and generally complicated designs were made, many types of moulding and other planes were used. In most cases these were of wood, with single irons, little different from similar planes today.

New ideas are still being tried, such as grooving the face to prevent drag, plastic handles, built-up handles, and hard wood facings, as on

Figure 7.15 Modern electric router

continental planes. The main improvement is in better and more even quality of steels for the cutting iron and the use of special tool steel for work on plastic laminates.

Multi-purpose planes, with a few exceptions such as rebate planes, have been superseded in the workshop by the use of portable power routers.

Drawknives and spokeshaves do not appear very often among finds of ancient tools. An iron drawknife was found at Pompeii and one at Auvergne, both dating from the late Roman period. One, with tangs bent up for handles, was found among Viking shipwrights' tools, dating from the tenth century. In medieval Russia, drawknives were used for smoothing timber but they do not appear in Western illustrations of the same era. *Smith's Key*, of 1816, gives a full list of tools for carpenters, wheelers and coopers, which includes the drawknife. In 1683, an illustration of Moxon's writings showed a modern form of drawknife among carpenters' tools.

The word 'spokeshave' is recorded in 1510, but the earliest illustration is shown in *Smith's Key*. The name denotes that it was originally used mainly by wheelwrights (for shaping spokes), but it now has a much wider use. In America it was developed as a metal tool.

Chisels

As with other cutting tools, chisels first appeared as stone celts and pieces of bone or horn, later in copper, bronze and iron. No doubt the forerunners of chisels were multipurpose celts used for skinning animals, digging, cutting various material and for cleaning off wood after the bulk had been charred away. Many long pieces of flint or other hard stone dating from the Neolithic Period have been found, their carefully ground edges indicating that they could have been chisels.

Copper and bronze celts, without handles and shaped like those of stone, appeared soon after in the Bronze Age. Increased knowledge of smelting and forging led to socketed types, and chisels and gouges with wooden handles have been found (Fig. 7.16). Tang types of the same period have also been discovered (Fig. 7.17). Whether the tang or the socket was developed first is not known, though some early examples with flanges, like palstaves, suggest precedence for the socket type.

The most surprising fact is that chisels similar in shape to modern tools developed so early in time. Bevels on both sides forming a cutting edge, similar to modern cold chisels, soon gave way to a bevel on one side only. Tapering the width of the blade appears to have been common practice, some blades being widened considerably at the cutting end. Metal gouges appear less frequently than those of stone. Early tools were forged and are not so well shaped as those made later by casting. Modern chisels are made by drop forging, which forms the shape in a mould or die, and are finished by grinding.

Many pictures and relief carvings show craftspeople using chisels of some kind, often with a mallet. Specimens have also been found, usually in association with other tools, in all parts of the world, particularly in European countries. Among important finds are bronze chisels of both tang and socket types dating from 900 BC at Bologna, Italy; and in Glastonbury lake village, England, an iron gouge dating from 200 BC, its socket in the form of a palstave and its turned wooden handle well preserved in the local peat. In this area of England, many examples of bronze and iron tools and ornaments have been found, though no cutting tools figure among the bronze specimens.

The Assyrians, who were expert smelters, used copper chisels as well as those of bronze and iron. We know from drawings that the Egyptians used chisels, often wide stones without handles. It was not until the expansion of Imperial Rome that the more modern types in copper, bronze and iron made their appearance. Tang and socket iron chisels and gouges dating from the Roman expansion, but very similar to tools of today, have been found in many places. Iron chisels (one with a ferrule) dating from AD 75 have been found at Pompeii.

As in the history of planes, there is a gap in our knowledge of chisels from early to medieval times. In Russia between the tenth and thirteenth centuries, cutting edges of steel were welded to solid and socket-type iron blades, which appear to have been made for wooden handles although none have been

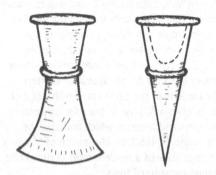

Figure 7.16 Cast bronze socketed chisels

Figure 7.17 Iron tanged chisel and ferrule, Thebes, Egypt, 700 BC

found (Fig. 7.18). The sixteenth century yields some specimens—in the Novaya Zemlya (Dutch) collection from 1597, a bevel-edged paring chisel with a shouldered tang and octagonal wooden handle is of particular interest. However, most information regarding chisels of the fifteenth, sixteenth and later centuries comes from descriptions, illustrations and carvings, which show chisels of all types, a few with ferruled handles, most of which are square or octagonal.

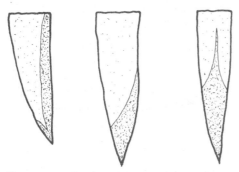

Figure 7.18 Steel cutting edges welded to iron chisel blades

Felibien, in a textbook published in 1676, illustrates both the straight blade, called 'ciseau', and the splayed type, called 'fermoir'. Moxon, writing in 1683, mentions these two types, stating that the ciseau, or paring chisel, is used without a mallet and explaining that the fermoir, or former (present-day 'firmer') is used 'before paring chisels as in the fore plaine' (Fig. 7.19). Peter Nicholson shows, in his textbook of 1812, a mortice chisel with elliptical-sectioned handle and a firmer chisel with round handle, both without ferrules. The elliptical-handled chisels often had a heavy shoulder, with leather buffer pieces at shoulder and top. A core driver, at least a hundred years old, has been found with an ash handle of this type, though handles were more often of beech or boxwood.

Modern chisels (Fig. 7.20) are made in both tang and socket types for all purposes. Turned wooden handles, ferruled at the bottom and

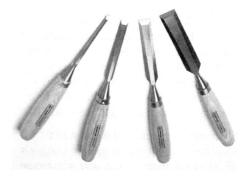

Figure 7.19 Seventeenth-century paring and firmer chisels

often at the top, are generally of two shapes—for firmer and paring chisels, a thin neck permits finger grip while on mortice chisels, the bulbous-shaped handle fits the hollow of the hand. Hexagonal handles of boxwood are occasionally seen on paring chisels. All types are now available with plastic handles moulded to shape for finger or hand grip, but the chief improvement in present-day chisels lies in the more even quality of the steel used for the cutting edge.

Figure 7.20 Modern tanged and socketed chisels

Saws

Authorities believe that saws were not invented and deliberately made as such, but gradually developed from chipped flint knives, whose rough edges, it was found, would cut if used with a sawing action (Fig. 7.21). These so-called saws were often backed with a wooden handle or had a rib along the side, which would have prevented sawing right through (Fig. 7.22). One interesting specimen from northern Italy, however, appears to have been made as a saw and consists of pieces of flint set in a wooden handle.

Figure 7.21 Stone Age flint saw

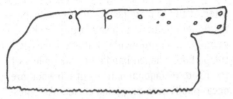

Figure 7.22 Roman backed saw

Flint saws dating from the times when Neanderthals lived on earth (from 130 000 to 30 000 years ago) have been found in caves in France, stoneheaps in Denmark and Sweden and lake dwellings in Switzerland and northern Italy. Flint tools similar to these European examples were discovered in New Mexico, while peoples in the Caribbean Sea area evidently used notched shells. Saws made from a type of volcanic glass, found in Ur in Mesopotamia, date from 5000 to 4000 BC.

Specimens discovered and writings and illustrations from the time confirm that tools designed and made as saws were in common use during the Copper and Bronze ages. Both the Greeks and the Egyptians have been credited with being the first users of metal saws, but the most authentic records come from Egyptian pictures. Biblical references to saws are numerous. In the fifth century BC, Aristophanes and Thucydides refer to saws and sawing. Sophocles (496–406 BC) writes of 'sawdust shed from a saw's teeth where men work with timber'. Pliny (AD 23–79) and other writers attribute the invention of the saw to Daedalus.

Many copper and bronze saws and the moulds (Fig. 7.23) for casting them have been found in places as widely dispersed as Egypt, Crete, France, Spain, Hungary, Italy and Sweden.

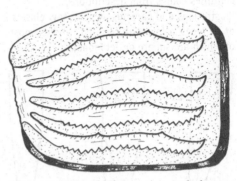

Figure 7.23 Mould for casting early bronze saws

Egyptian copper saws dating from 1490 BC, one with a blade 500 mm long, have wooden handles and are shaped like modern bread-knives (Fig. 7.24). Similar tools are pictured on the coffin of Sopi, dating from 2000 BC (Fig. 7.25). The teeth of early Egyptian saws, which were without set, sloped toward the handle and cut on the pull stroke—a model of a carpenter's shop of 2000 BC, now in the Cairo Museum, shows this type of saw in use. Saw marks on wood from a tomb 5000 years old indicate that cuts were made from each corner.

At Crete a number of saws dating from the Minoan period show advanced design. Three large blades—1680, 1590, and 1575 mm long respectively—were shaped and tapered, with holes for securing a handle. The longest saw had uniformly cut teeth of 4 to 5 points per 25 mm; the next in length had 7 points

Figure 7.24 Egyptian copper saw, 1490 BC

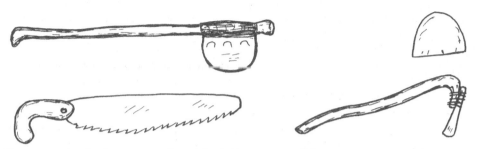

Figure 7.25 Tools shown on coffin of Sopi, Egypt, 2000 BC

per 25 mm near the wider end, 8 points per 25 mm near the toe and 4 points per 25 mm for the remainder; the shortest had 8 points per 25 mm near the toe and 5 points per 25 mm for the remainder. Another Cretan saw, 100 mm long and shaped to 70 mm wide at the square end, 100 mm at the centre and 38 mm at the toe, had small, irregular teeth cut on a curved edge. No handle was found with the saw, but holes in the square end were apparently intended for attaching a handle.

Bronze saws have been discovered in England, Italy and Sicily, and one from Russia is unique in that it has a socket to take the handle, and teeth on both edges.

Iron saws (Fig. 7.26) have come from the La Tene site in Switzerland and similar examples from the Glastonbury lake village, England.

Figure 7.26 Early Iron Age saw

It is uncertain when saws were first set, but the technique is believed to have been developed in Italy. Pliny refers to the set of saws in his *Natural History*, saying that green timber is more difficult to saw than moderately seasoned timber and 'it is for this reason that the teeth of the saw are often made to project right and left in turns, a method by which sawdust is discharged'. A saw found at Glastonbury, dating from before

Roman expansion, had 68 teeth bent alternately from side to side.

The problem of preventing saws from buckling due to being forced accounts for most of the specimens found having teeth raked towards the handle. Other means of preventing buckling were evidently tried— some saws were given teeth of various sizes; some were provided with a rib; others had the handle extended along the top, thus possibly paving the way for modern back-saws such as tenon saws. The most effective way of preventing buckling is to hold the blade in a frame, keeping it in tension, and there is plenty of evidence that this was done. The Egyptians and Romans used double saws strained with a cord, which seem to have been the forerunners of the framesaw, and Roman relief carvings, paintings and pottery decorations often show framesaws in use.

As with other tools, there appears to have been little development in the saw from the time of the Romans until the fourteenth or fifteenth century. Several saws from Russia from about the tenth or eleventh century are of the bread-knife shape (Fig. 7.27) observed to have been in use elsewhere two and a half centuries earlier.

Figure 7.27 Eleventh-century Russian handsaw

It was not until a method was devised of preparing steel in wide strips that much change took place. Better shaped and evenly spread teeth appear in saws of the seventeenth century, and drawings of framesaws appear frequently during this period and earlier (Figs 7.28, 7.29). These include a number of frames very much like modern bowsaw frames, with string and toggle to achieve tension. In all of these illustrations the blade is shown in the same position—width of blade in line with frame. One picture shows turned handles, but they do not appear to be adjustable, and it is hard to determine exactly when the idea of turning the blade was introduced.

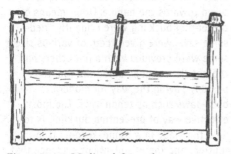

Figure 7.28 Medieval framed saw

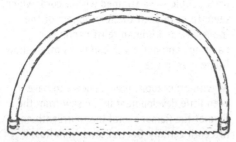

Figure 7.29 Russian saw in bent wooden frame, tenth to eleventh centuries

A fretsaw, dating from 1671, pictured on a silver tankard in the Stockholm Joiners Guild, is identical in shape to the modern fretsaw, having a thumbscrew and a round handle. From a shipwreck off the Danish coast in 1658 comes part of a frame with handles, similar to a modern bowsaw frame. In Europe, and less

often in England, blades were fitted into the underside of a wooden handle shaped like a German jack plane, and these came to be known as 'staircase saws'. Closed handles appear to be a fairly recent development (Fig. 7.30).

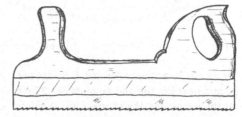

Figure 7.30 Swedish grooving saw, 1771

Framesaws and two-handled crosscut saws were developed for cutting logs, flitches and baulks into planks. They required two operators, one of whom worked in a pit below the log, and therefore these saws were known in England as 'pit saws'. In other parts of the world until the twentieth century, large wooden trestles were used to hold the log, and this method may still be adopted in remote areas. Catalogues from early nineteenth century toolmakers contain illustrations of pitsaws and framesaws with their special fittings, while dating from the Pompeii eruption of AD 79 onwards, many pictures illustrating the conversion of timber show logs supported on two trestles.

With the introduction of water-driven mills for rolling steel, wide-bladed saws came into use. This development was confined to England and Holland and each country evolved typical patterns, particularly as regards the shape of the handle and the method of fixing it. Dutch sawmakers continued to make handles of the pistol-grip type, but English makers developed an entirely new shape, the flat handle riveted to the blade.

The reason for the bead or nib often found at or near the toe of saws prior to the twentieth century has never been satisfactorily explained. Originally it may have been merely a hole intended for hanging the saw or, more probably, it was purely a form of decoration which later became a grade or quality mark similar in function to the incised designs on the handles of modern Disston saws.

A sabotmaker's saw in a museum in southern Belgium has the blade enlarged at the toe end and pierced for fixing in a frame with a peg. The whole saw is a frame like a narrow bowsaw, one end of the string being held in a curled notch on top of the handle—this may explain the curl in the top of some other saw handles.

Development in handsaws in the nineteenth century was mainly in the standardisation of types for particular use. The shape of blade and handle, length and width of blade and number of points per inch were carefully studied by manufacturers, who produced illustrated catalogues listing new and improved features of their products. Saw backs were straight until Disstons introduced the hollow-backed, or skewback, type in 1874, and most saws are now of this type (see the Chapter 'Care and Maintenance of Tools', in *Woodworking, Part One*).

Large crosscut saws for two-person operation were also developed along similar lines. As most of these were for felling trees or converting logs into timber, new tooth shapes and groupings were devised to suit the industry's needs in cutting timber of various kinds.

A glance at saw displays in hardware stores will reveal the wide variety of saws available today, yet basically the shape is the same as it has been for the last hundred years. The chief advances have been in the development of saws specially designed for cutting new, harder materials; plastic handles of simplified shape; sets of saws for a variety of work; and, as with plane irons and chisels, steel of more even quality (Fig. 7.31).

Figure 7.31 Battery powered circular saw

BORING TOOLS

A study of ancient boring tools shows that the method of holding and turning them presented a greater problem than producing the actual bit.

The earliest stone awls, or piercers, were without handles and were evidently twisted or turned by hand. Later when hafted pieces of flint, bone, horn or metal were used, it seems that the handle was rotated between the hands after the manner of a firestick; however, thongs or bows for turning appeared at a very early stage and pump drills, a clever development of the bow, were also used (Fig. 7.32). The transition from simple turning by hand to the strap or bow drill appears to have been abrupt, with no evidence of intermediate methods. By the time copper, bronze and iron were used for toolmaking, strap and bow drills, many of them two-person implements, were in common use in all countries.

Illustrations and specimens of boring tools date from 3000 BC in Egypt, all illustrations showing some form of bow drill. Boring holes was an important operation for early Egyptian woodworkers, for they used pins, dowels or lashing for most joining.

Greek and Roman illustrations also show bow drills, Roman shipbuilders apparently being the first to use a curved bow. In the Iliad,

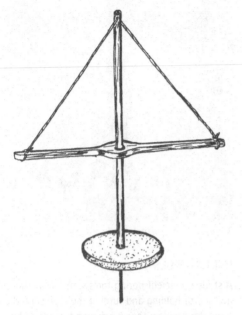

Figure 7.32 Stone Age pump drill, still used today by Eskimos and natives of New Guinea

Homer mentions boring 'with a drill while his fellows below spin it with a strap which they hold at each end'. After the Roman period, bow drills seem to disappear, to be replaced by the more powerful breast drill.

Early bits and drills were made with the cutting and handle parts fixed together, the separate bit and handle representing a much later development. The action was one of scraping rather than cutting, although a centre bit from Roman times has been found. The cutting part of the bit was usually flat with sharp edges (Fig. 7.33), similar in style and shape to a small bit which can be made by flattening and sharpening a nail. This shape was later improved and gouge, spoon, quill, and nose bits were evolved, some of which types are still in use. A specimen found at the site of a Roman encampment resembles a modern flat, high-speed bit.

No means of removing waste was devised until the invention of the auger bit by Phineas

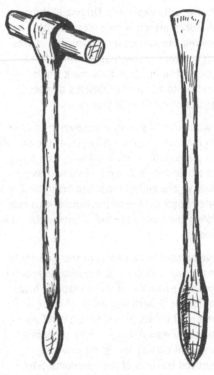

Figure 7.33 Russian augers, tenth to thirteenth centuries

Cook in 1770. Bits of similar pattern have been made until recently, differing from Cook's in having a draw thread to pull the auger into the wood.

Cross-handles similar to those of present-day gimlets (Fig. 7.34) have been used since Roman times. Many spoon-shaped iron bits may have had handles of this type, though none has been found in association with these tools. Pliny tells us that handles were made of olive, box, oak, elm and ash, while a gimlet found at Shrewsbury, England, had a bone handle. For larger bits, the turning handle was similar to the handle of a modern auger without the eye. Augers with a hole for attaching the handle appeared in the tenth century. Early augers, called 'pod augers', were like cobra bits with an extra twist, much taper and no draw thread.

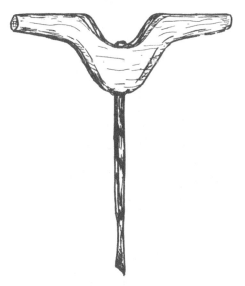

Figure 7.34 Medieval auger used straight spoon or shell bit

Several ancient Roman bits have been found which would fit into a brace, though no braces were found with them. Early braces were chiefly of wood (Fig. 7.35), metal braces made subsequently achieving little popularity. The first wooden brace appeared in the fifteenth century, with a square socket into which the bit fitted without being secured in any way. Later methods of holding the bit in the brace included bifurcating the bit stock so that the split end projected through the socket and opened out, fastening with a pin through both stock and socket, and securing by means of a thumb screw.

Development of the brace was fairly rapid from the eighteenth century onwards.

Figure 7.35 Brace-and-spoon bit, Amsterdam, Holland, 1596

Improvements included strengthening the crank with brass plates (Fig. 7.36), provision of a metal rotating head and, in the nineteenth century, a free handle on the crank and extensive modifications to the chuck.

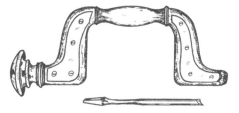

Figure 7.36 Framed brace by William Marples, 1860

A spring-button chuck was patented by William Marples in England in the early nineteenth century. In 1859, the Spofford split-socket chuck with thumb screw was patented. The first chuck with adjustable jaws was patented in 1864 and was later given interlocking jaws. The next development was the Barber brace chuck followed by the Fray brace chuck. In 1868, a patent was taken out in the United States for a ratchet brace, but this did not become popular for many years.

The most recent working improvement, as distinct from refinements such as the plating of metals parts, has been the installation of ball-bearings in head, handle and chuck (Fig. 7.37).

All modern drills are equipped with efficient chucks operated with a key. Breast and power drills may be used for boring large and small holes, while useful tools for small holes are

Figure 7.37 Modern ratchet brace

the hand drill and the push drill operating on a spiral, a development of Archimedean spiral drills (Fig. 7.38).

(a)

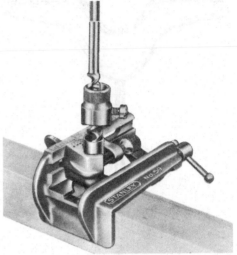

Figure 7.40 Boring jig for dowelling, note depth stop on dowel bit

(b)

Figure 7.38 (a) Modern hand drill and (b) Archimedean drill

Figure 7.39 Hand drill with enclosed gears

DEVELOPMENT OF WOODWORKING MACHINES

Although wood has always been one of the chief materials used by people, it was not until the fifteenth century that woodworking machines of any significance were invented. Even then very little progress was made in making machines until the end of the eighteenth century.

The Industrial Revolution, which took place during the latter half of the eighteenth century and the first half of the nineteenth century, introduced machines into crafts that were traditionally hand and home crafts.

It is interesting to note that it was the reform of the English prison system in about 1790 and the subsequent introduction of workshops equipped with hand tools into prisons that created a special demand for woodworking machines. Wood was the principal material to be used, yet very few prisoners had any trade skills and, without machines, little use could be made of the workshops. Sir Jeremy Bentham, who was mainly responsible for the reform, called on his inventive brother Samuel to solve the problem by inventing machines. Samuel Bentham was well qualified to do this as he was a trained shipbuilder and naval engineer with a number of inventions to his credit. He had been sent to Europe by the government in 1779 to gain further experience in these areas. On his return to England in 1791 he was appointed inspector general of dockyards. The machines he invented for use in prison workshops were so successful that they were introduced into dockyards.

The two main problems that confronted early inventors were the lack of power to drive the machines and the scarcity of suitable materials to make them. Later other problems arose, such as those connected to the lack of bearings and lubrication. It must be remembered that many problems became apparent only after machines had been in use because the inventors had no previous experience to guide them. Machine design and invention is a process of evolution with improvements and developments going on all the time.

POWER

Power is not a problem with modern machines as a wide range of electric and combustion motors are readily available, one or more for each machine. Some high-speed machines use compressed air turbines.

Early machines were driven by people, animals, wind or water. Human labour was used to turn machines by a hand-operated crank and wheel or with a treadle. Obviously the power was

Fig 7.41 Modern power tools

limited. The main driving wheel, usually large in diameter, was made of heavy materials to act as a flywheel. This helped to increase speed and carry over momentum. Variation in the size of pulleys or wooden cogs was used to obtain the necessary speed.

One of the earliest methods used to produce rotary motion was the springy pole, as used on a pole lathe. From the pole lathe came the idea of using a pulley or wheel worked by a treadle and connected by some form of belt to the job or a pulley on a spindle. This gave continuous rotary motion in one direction.

Horses or donkeys working a form of treadmill or walking around and turning a shaft gave more power but had disadvantages, space being the main one. The term for rating power—'horsepower'—is derived from the use of horses. The term 'donkey engine' obviously came about in the same way.

Wind provided plenty of power but there could be no certainty as to its availability and force. Windmills are still used, particularly on pumps to draw up artesian water. Water-power could be made almost constant as to availability and force, but required the workshop or factory to be near a running stream.

Steam-engines, invented in 1769, were used but it was not until the middle of the nineteenth century that Corliss developed an economical, practical steam-engine.

Power from one source could be used to drive a number of machines by means of a complicated system of belts running over wheels on overhead shafts or countershafts. This system was wasteful of power and difficult to make safe. Even after the introduction of electric and combustion power, only one power unit was used with a belt system for many years. Power could be diverted from any machine by shifting the belt from the fixed wheel to a loose pulley wheel. In factories with many machines there was a maze of shafts, pulleys and belts.

With the introduction of electric or petrol motors for each machine, safety measures were improved and the loss of power greatly reduced.

An important development in the use of electric motors was the introduction of frequency changers. These are electrical devices which increase the alternating current impulses. By increasing the number of cycles the speed is increased. A motor running at 3600 rpm with 60 cycles will double its speed if the number of cycles is doubled.

Two very important features of electric motors are that they run quietly and produce no exhaust noise or fumes. Engine noise and exhaust fumes can create major pollution problems in industrial areas.

MATERIALS

Wood

Up to the end of the eighteenth century, wood was the main structural material. Bridges, ships, vehicles, buildings and furniture were all made of wood and craftspeople and designers thought out their ideas in terms of what could be done with wood. The frames and bases of machines were of wood while metals were used for shafts, bearings and cutters. Because wood swells and shrinks with changes of air conditions, it is difficult to build a constantly rigid frame or base. This allows vibration which affects the finish of the work being done and causes much wear on moving parts of machines. Large timber sections were necessary for strength, but this was often impracticable.

Metal

With the introduction of improved methods of producing and working metal in the latter half of the nineteenth century, metal was used for all parts of machines. In modern machines a wide range of metals, alloys and plastics are used. Improved methods of casting and welding have also helped with designs.

SAFETY

Early machines were dangerous to use because little thought was given to safety precautions. Belts, cutters and gears were not guarded and cutter heads were square.

Modern machines are subject to strict safety regulations regarding their design, installation and operation. Inspectors regularly visit factories and workshops to enforce these regulations.

BEARINGS

For machines running at high speed or under heavy loads, efficient bearings are most important.

Bearings must be true, close-fitting and strong enough to carry the power applied. Bearings on early machines were of two kinds—plain metal blocks with a hole for the shaft, or spindle and cones. Both kinds were unsatisfactory at high speed or under a heavy load.

The metal blocks were bolted to the base or frame. Later, bushes or linings were fitted into them. A further improvement was made by making bearings in two parts, the top half being bolted through the bottom half, over the shaft. This type allowed for some adjustment and was easier to reline. It was soon found that the use of the same kind of metal for both spindle and bearing was unsatisfactory, and softer metals were then used for bushes or liners. Special alloys of copper, antimony and tin were found to be suitable. Isaac Babbitt (1799–1862) invented a special alloy called Babbitt metal. This type of metal is still used for liners, and bearings are said to be Babbitted.

Cone bearings held the spindle or shaft between centres. The supports and spindle had a conical point and conical hole which fitted into each other. This type gave easy adjustment but wore rather quickly.

The introduction of ball- and roller-bearings was a big improvement as friction was greatly reduced. Further developments in the use of newer materials such as nylon and teflon are being made.

Lubrication

Proper lubrication of bearings is most important. To enable bearings of early machines to be lubricated easily, a small hole was made in the top. Sometimes small cups or containers to hold oil were fitted into this hole and kept filled so that lubrication was continuous. This method is still used, but new lubricants enable the use of sealed bearings which require no service during the normal life of the bearing.

Until the discovery of mineral oil, which is non-drying, effective lubrication was a problem as other kinds of lubricants dry out. Modern lubricating oils and greases are highly specialised products, with a variety of types and grades to suit all requirements.

MACHINES

Lathes

Illustrated in Figure 7.43 are some important stages in the evolution of the wood-turning lathe.

That lathes existed in some form at least as early as 1400 BC is evidenced by Egyptian relics of stool legs and the feet of bedsteads made in this era and obviously shaped by turning. The earliest known picture of a lathe, on a tomb relief of about 300 BC, shows how the work was rotated by means of a strap.

Turned bowls and wheel hubs found in English peat-bogs are believed to have been produced soon after the Celtic people of the Iron Age introduced the lathe to England in about 200 BC.

During the early Middle Ages, a more powerful and convenient method of turning the work was contrived in the pole lathe. Sometimes the pole was fixed in a vertical position so that the strap passed over a loose pulley and down to the work and the treadle. A similar form of lathe was used in the nineteenth century in England by the bodgers (woodcarvers) to turn chair legs from the green wood of the forests.

It is interesting to note that the design of most lathes of Roman and medieval times included a long bench for the operator, who, in those days, was required to work very long hours, usually from dawn to dusk.

All these early lathes had the great disadvantage of an idle return stroke. One of the first machines with a continuously rotating drive was illustrated, if not invented, by Leonardo da Vinci in the year 1500.

Sketches drawn by Moxon in 1683 illustrated the great wheel lathe and the treadle and flywheel lathe in many stages of development. Stepped and grooved pulleys provided a better range of speeds, and there was even such sophisticated equipment as drilling attachments.

Availability of steam, water and electric motive power greatly assisted the rapid developments in lathe design in the nineteenth and early twentieth centuries, culminating in the modern computer-controlled, variable-speed lathes

Figure 7.42 A treadle wood-turning lathe from 1884 with fretsaw attachment: this machine was available with all accessories, including emery wheel shown, at total cost of £2:14:0 (approximately $4.00)

used in industry today. The 'Leady' lathe shown in Figure 7.44 features a number of innovations to assist the wood-turner. It is portable and self-stabilising; all adjustments are by cam action; it has visible speed, automatic belt tension and integrated dust extraction; and it converts easily from a bowl to long bed lathe as required.

Recent innovations to lathe design include the use of both swivel and sliding headstocks; the swivel head allows the operator to face the job instead of having to reach across the lathe; and the sliding head does away with the need for an outrigger or the need to turn off-end.

Machine saws

Cutting up logs or large sections of timber is a slow job if done by hand and some of the first machines invented were for this purpose (Fig. 7.46).

Fifteenth and sixteenth century documents record the use of reciprocating framesaws in England, Germany, Holland, Norway and Sweden. These were direct descendants of the pitsaw. Framesaws are still used today.

The first circular saw was invented in Holland. It was not a success because of the difficulty of making saw blades and bearings at that time. In 1777, Samuel Miller, a sailmaker at Southampton, England, patented a saw of similar type.

Early in the nineteenth century a number of manufacturers were making woodworking machine saws. Once machines were made and put into use, new ideas and improvements rapidly followed. These included tilting saw, rising and falling table, rising and falling saw, conical saw arbor, cut-off gauge, splitting gauge and extension of top or bench. Frames and tops were made of wood until about 1850 when cast iron was used.

Circular saws were originally designed to be mounted in a bench and used mainly for

THE STORY OF THE LATHE

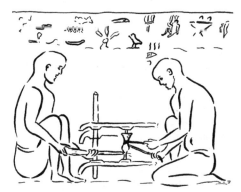

Egyptian wood turners at work in the fourth century B.C. A wall scene in the tomb of Petosiris.

Egyptian bow lathe. Still used in primitive countries.

Early tree lathe.

Early crank and flywheel lathe by Leonardo da Vinci, 1500.

Pole lathe—first used during the early middle ages.

Great wheel lathe, Moxon 1683. Later improved to treadle and flywheel using stepped pulleys. Water, steam and electric power followed.

Figure 7.43 Development of wood-turning lathe

Figure 7.44 Modern Australian-designed combination lathe has a 500 mm swing and one metre between centres. Bowl lathe converts to a long bed lathe when swing bed is raised

ripping, but modern saws have a much wider range (see pp. 39–41). There are three main kinds of circular saw besides those mounted on a bench—pendulum, radial arm and portable.

Pendulum or swing saws are mounted on an arm pivoted at the top, like a clock pendulum, so that the saw can be moved across the timber.

Radial saws are more versatile in that the arm can be moved through a wide arc. This arm supports the saw and power unit, which can be moved backwards and forwards, and tilted. Cutting may be done at any angle to the side or edge by using the swing of the arm and/or tilting the saw. Special blades can be used for cutting trenches or grooves.

Gang saws, that is, two or more saws mounted on one spindle, are used for making a number of cuts simultaneously.

For breaking down very big logs in sawmills two saws are mounted, one above the other, in such a position that their cuts will overlap (see *Woodworking, Part One*). This enables a cut to be made almost twice the radius of one saw.

For trimming such materials as plastic laminates, particle board and hardboard to length or width, two saws running in a guide along each side of a bench ensure parallel cuts. In factories where these materials are used in quantity, singly mounted saws are used to cut the sheets to the required size. The saws operate in horizontal or vertical frames.

Chain saws do not have the conventional blade, but have cutters mounted on a chain that runs around a steel blade. The chain is driven by an electric motor or petrol engine, the whole saw being portable. These saws are mainly used for felling trees, trimming branches and cutting logs or heavy timber to length.

Hand-held power circular saws and saw attachments are illustrated in Chapter 2.

Bandsaws were invented in 1808 by William Newberry. Early bandsaws must have been dangerous to use as both wheels and the blades were left unguarded. Because of the difficulty of joining the blade, little progress was made in the development of the bandsaw until fifty years later, when the quality of saw blades was improved in France. A machinery catalogue of 1884 describes and shows a bandsaw. The frame is of cast iron with a tilting table. The 965 mm wheels have wrought-iron spokes and bentwood rims, are rubber faced and run in long adjustable bearings; the upper wheel has a tilting device so that the saw can be made to run true over the wheels. The machine had a swing of 940 mm and 330 mm depth of cut. Operating speed was 370 rpm. Tight and loose pulleys were fitted so that the machine could be connected by a belt to a power source.

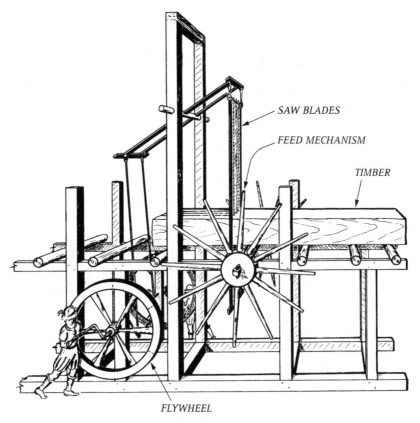

SAW BLADES

FEED MECHANISM

TIMBER

FLYWHEEL

Figure 7.45 Ancient sawmill used in fifteenth and sixteenth centuries

In general appearance the first bandsaws were similar to those of today, except that the upper and lower wheel guards were not introduced until about 1910. The main improvements have been in the saw blades—in the method of joining the blades to the saw, in their flexibility and the shape of their teeth. Special blades for sawing metal and plastic materials have also been developed. Originally handsaws were designed with narrow blades for cutting curves. Later, fences or guides were introduced, so that handsaws became more general purpose machines.

Jigsaws are also called scroll-saws (Fig. 7.48). These have been in use since the beginnings of the Industrial Revolution and operate with a short reciprocal action. Although they can be used for straight cutting they are used particularly for cutting intricate shapes, hence the term 'jigsaw puzzle'. Some of the early saws had a frame in two parts, the top arm being pivoted and spring loaded to keep the saw in tension. Modern saws have a one-piece frame with a spring tension attachment in the end of the top arm. The reciprocal motion is made by a crank below the table. Saw blades are held by a clamp at each end. Power is usually provided by a 450 watt to 900 watt electric motor. Wood up to 50 mm thick can be cut.

Planing machines

In 1776, a patent was taken out for a planing machine by a man named Hutton. It worked by a series of cutting irons being dragged over the board by means of pulleys. The idea had no real value and was never developed.

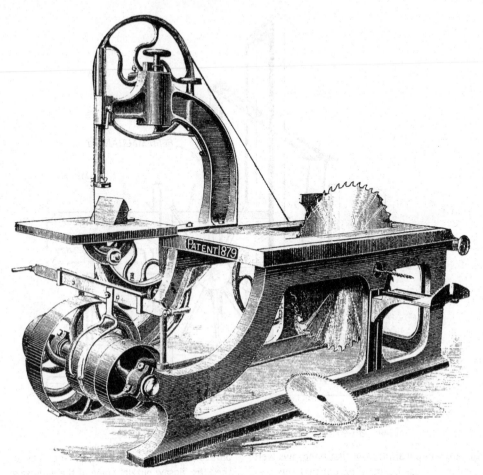

Figure 7.46 A combination machine adapted for steam power; it was a bandsaw, circular saw or boring machine as required and sold in 1879 for £48:00:0 (approximately $96.00)

After his return to England in 1791, Samuel Bentham invented a number of machines, the most important of all being the rotary cutter. The principle of rotary cutters—the cutter works through a slot in the table—is used in many modern machines. At first these rotary cutters consisted of steel blades bolted on to the sides of a square block. This left gaps in the opening in the top, which was dangerous. In about 1901, the cylindrical cutter head was developed in England and Germany, and in 1908 the Oliver Machinery Co. introduced it in the United States.

In 1802, Joseph Bramah invented a transverse planing machine, which operated in much the same way as modern machines planing metal. Cramped to a carriage, the board moved under horizontal revolving cutters which could be changed to produce various sectional shapes in the timber. This machine was later improved and soon many were in use.

As time went on improvements made in the design produced a planing machine with better bearings, safety features, steel rather than wood tops, roller feed, better steel for

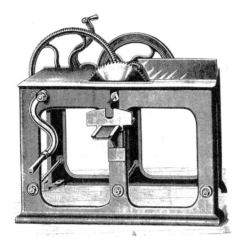

Fig 7.47 An 1888 hand-power circular saw bench was for use when steam was not available; fitted with gunmetal bearings, canting fence and a 9 inch blade; price then was £16:00:0 (approximately $32.00)

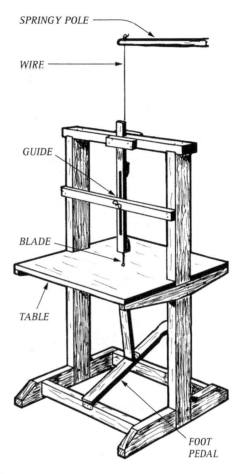

Figure 7.48 An early type of jigsaw with reciprocating cutting action produced by foot treadle; blade was tensioned by wooden rod or pole attached to ceiling

cutters, and cutter heads above and below the table and for the edges, so that up to four cuts could be made simultaneously. Machines with a single cutter below the table are called jointers. Machines with a cutter above an adjustable table are known as thicknessers. These machines may also have another cutter below the table so that both sides are dressed in the one operation. For dressing both sides and edges, machines with four cutting heads are used. Early machines had belt drives to each head from the one power unit. Modern machines use power with belt or direct drive for each head.

Sanding machines

Sanding machines operate an abrasive disc, belt, drum and pad. Originally, finishing abrasives were made by coating paper with natural sand, hence the term 'sandpaper'. Although sand is not now used, these machines are called sanders. The development of sanders is more recent than that of other machines because suitable abrasives were not available until the end of the nineteenth century. Finding satisfactory methods of joining abrasive belts was another problem.

Modern sanding machines are very efficient and have been developed to finish straight and shaped wood. There are many types, each being designed for particular operations.

Belt sanders are the most versatile of the sanding machines and consist of a cloth-backed abrasive running over two or more pulleys, powered with direct drive motors. The straight part of the belt may be

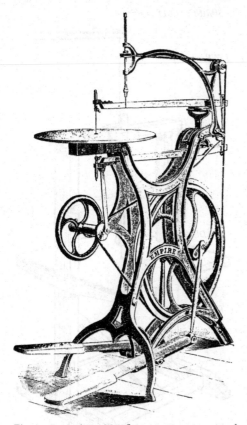

Figure 7.49 An 1884 foot or steam-powered jigsaw or scroll-saw, designed for carpenters or cabinetmakers; allowed for 60 mm swing and 75 mm thickness of timber, price was £6:00:0 (approximately $12.60)

shaped work narrow belts are used, guided by wheels to run over the work which ensures that the belt follows the contours. The work can be turned or oscillated to bring it into contact with the belt. Difficult shapes may require several operations to completely finish all parts.

Wide belt sanding has been recently developed using belts up to 1820 mm wide. The belts have a long life while the machines produce a fine quality finish and have a high rate of production. Output rate is up to 30 metres per minute with belt speed up to 1700 metres per minute. Narrower belts usually run at about 1000 metres per minute.

Linishers are types of belt sanders in which the belt runs over a table, the work being pressed on to the belt (Fig. 1.51). Flat or shaped work may be sanded on the flat position or, if concave, on the pulley. Tension is controlled by adjusting one of the pulleys or by the use of a third pulley.

Edge sanders are similar in operation to linishers except that the pulley axes are vertical. The belt runs over a vertical back plate. To keep the abrasive clean and lessen friction both belt and back plate oscillate.

Drum sanders have an abrasive fitted round a cylinder or drum. These machines range from a single drum on which the work may be held by hand to three or more drums mounted in a frame with power feed. Different grades of abrasives are used on the drums so that all operations from roughing to finishing are done on the one machine. Two drums, one above the other, are used to sand both sides of the work in one operation. Where two or more drums are used, the lower one and the feed table can be raised or lowered for different thicknesses of work.

To prevent clogging of abrasives with wood dust, the drums oscillate slightly. A suction system is also used to pick up the dust.

horizontal or vertical. Machines operating up to three separate belts with different grades of abrasive complete all sanding operations from roughing to finishing.

The frame to hold the work is mounted either below the lower part of the belt or between the two parts and moves backwards and forwards so that the full width of the work can be traversed. Generally a hand pad is used to press the belt on to the work. Machines for production work have power-operated pads or a contact roller. Tension on the belt is usually adjusted by an idler pulley. For finishing

Drums for sanding small jobs can be made with any diameter and used between centres of a lathe or mounted on a spindle for use in a hand drill, lathe or flexible shaft. Drums can be cylindrical or conical to suit the work and are made of wood, metal or soft rubber. The abrasive is either glued on or wrapped around and held by a wedge-shaped slip.

Pneumatic rubber drums covered with an abrasive sleeve are used for sanding complex curved shapes.

Disc sanders have an abrasive paper or cloth glued on to a disc fitted to a spindle or screwed on to the face-plate of a lathe. Machines with direct power operating a large disc are also used (Fig. 1.54).

Discs may be used on portable machines, such as drills, but are usually in a fixed position on an adjustable work-table. Tables can have attachments fitted for sanding mitres or the edges of circular work. Slow speeds give best results as the abrasive wears quickly or clogs if used too fast. A few trials will soon determine the speed for best results. Work is held on the table on the side of the disc running down to it. Discs can be used for finishing end-grain, straight or convex work. If the full circle of the disc is used, some of the abrasion is across the grain.

Orbital sanders move backwards and forwards as well as sideways. For this reason some of the sanding is across the grain, so that a high finish is not possible. An orbital sander is made as a complete unit with its own power or as an attachment to a drill (see Fig. 2.13).

Reciprocal sanders are similar to orbital sanders but do not move sideways. They give a better finish than orbital sanders but are rather slow (see Fig. 2.13).

Portable sanders are made in disc and belt types. The disc has a flexible rubber back which allows it to be used for curved work. It is also used for sanding floors. The belt variety has a back or platen similar to a linisher. It is used for sanding work on the bench, floor sanding and finishing edges or surfaces which could not be done on a fixed machine (see Fig. 2.9).

Automatic sanders are used only for mass production work as they take a long time to set up. The work to be sanded must all be the same size and shape. Work is placed in a hopper and picked up automatically between centres. It is rotated against an abrasive belt and fed into a hopper after a series of operations.

More information on sanders can be found in Chapter 1.

AUTOMATIC MACHINES

A further step in machine design was the introduction of semi-automatic and automatic machines. Semi-automatic machines usually require the work to be set up and removed from the machine by hand, the actual work on the material being done automatically. An automatic machine can set up the work, carry out several operations and remove the finished article. Some automatic machines are operated electronically and require very little attention once the correct sequence of operation has been set up. Setting up takes considerable time so these machines are used only for mass production.

In 1819, Thomas Blanchard of Philadelphia invented a semi-automatic lathe. It was really a type of copying lathe, as a replica of the job to be turned acted as a guide for the cutters.

Copying lathes are examples of semi-automatic machines. Work has to be set up and removed from the machine, but the shaping is done automatically. These lathes reproduce one or more copies of a prepared shape or copy. Arms transfer the shape of the copy to the cutters which simultaneously shape one or more pieces of timber held between centres. Both job and cutters can be rotated and the cutters move along the job. Wooden heads for golf clubs, furniture legs and chair arms are examples of work done by a copying lathe. Generally copying lathes are used for mass production.

A recent development for small factories and home workshops is multipurpose machines. By using attachments, they can be adapted to do a variety of operations using the same power unit. Another use of attachments is with a power drill for sawing, drilling, routing and sanding.

FURNITURE

The study of the woodworker's craft in furniture design and construction can take us back little further in time than the fifteenth century. Prior to that, furniture was scanty even in the houses of the well-to-do. Stools, forms and trestles provided seats and tables, chairs being reserved for the head of the household (hence the modern word 'chair' used to describe the convenor of a meeting). Chests were used for storage and the larger ones as beds. The peasantry, living in fortified settlements and serving their feudal lords in return for protection and mere shelter, did not have the wherewithal to furnish and beautify their homes. Nor did they have the incentive, subject as they were to the constant threat of attack, pillage and burning by the warring feudal armies.

Travel and transport were slow and hazardous so there was little exchange of ideas or materials between geographic areas. A profound influence was exercised by the Church, a powerful self-supporting body which employed its own craftspeople, among them carpenters who made furniture conforming with the vertical, upward-thrusting lines of the prevailing Gothic style of architecture, the only style they knew.

By the end of the fifteenth century, the feudal system had collapsed and the Church's influence had begun to wane. The expansion of communications and voyages of exploration and discovery provided a wider knowledge of fresh ideas and new materials, which brought about great changes in furniture styles. These changes were at no time precise and clear-cut, but subject to considerable overlapping from the point of view of design, types of furniture produced and kinds of materials used.

ENGLISH PERIOD FURNITURE

In England, furniture styles can be grouped historically into four ages:

1. *the age of the carpenter, 1500 to 1660;*

2. *the age of the cabinet-maker, 1660 to 1750;*

3. *the age of the designers, 1750 to 1800;*

4. *the age of the machine, 1850 onwards.*

Roughly parallel to these chronologically are other groupings determined by the kinds of timber and materials most widely used at the time:

1. *the age of oak, to 1660;*

2. *the age of walnut, 1660 to 1725;*

3. *the age of mahogany, followed by satinwood and rosewood, 1725 to 1800;*

4. *the age of wide variety in materials and innovations such as plywood, made-up building boards, plastic and new adhesives, extending into the twentieth century.*

The period furniture styles into which the ages subdivide, and the preceding Gothic style, are discussed below. The principal characteristics of each are related to chairs, tables and chests, which are the items of furniture reflecting most clearly the nature of developments in style.

Gothic (eleventh to early fifteenth centuries)

Oak was the principal timber used in Gothic constructions, since it was plentifully available.

LEG STYLES

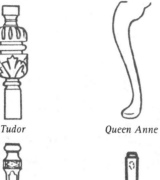
Tudor

Queen Anne

Chinese Chippendale

Chippendale

Sheraton

Hepplewhite

Adam Brothers

Jacobean

CHAIR STYLES

Adam Brothers

Chippendale

Hepplewhite

Sheraton

Colonial

Victorian

Regency

Figure 7.50 Comparison of typical leg styles and chair styles in England period furniture

Chairs had high, straight backs decorated with carving, straight legs with low rails between, filled-in front and sides, the latter carried up to form armrests. Benches had solid or panelled ends, and backs the same height or higher, sometimes with a canopy or ornamented with a finial.

Tables evolved from trestles and were similarly long and narrow. In monasteries it was customary to serve a light meal, or 'refection', for travellers. The meal room was called a 'refectory', hence 'refectory table', a term still in use.

Chests, originally crudely hollowed from logs, were later made from split timber held together with wooden pegs. They had no legs and the lid was hinged (Fig. 7.52). From chests were developed the cupboard, the cabinet, the buffet and the wardrobe.

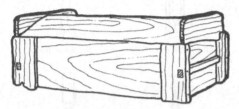

Figure 7.51 Fifteenth-century chest with pin-hinged lid

Tudor (1485–1558)

Gothic designs and construction were still used during the early part of the sixteenth century (Fig. 7.53) until artists and sculptors from Italy brought with them the Renaissance styles.

Chairs, now in wider use, had lower backs and were generally more comfortable. The transition was from the panelled box design, known as 'wainscot', to that with bulbous turned legs, rails at or near the foot and carved or panelled backs. Cushions were introduced.

Tables were still virtually trestles but had solid carved sides with a horizontal stretcher between, heavy bulbous turned legs, often carved, and plain rails right round at floor level. Top rails were heavily carved and supported on carved shaped brackets, while the edge of the top was moulded and often also carved.

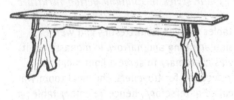

Figure 7.52 Early sixteenth-century long form or stool with Gothic shapings

Chests with drawers were introduced and were called 'mule chests'. They had straight, square legs, often decorated and forming part of an underframe (Fig. 7.54).

Fig 7.53 Mid sixteenth-century settle is simply a storage chest with back and arms

Elizabethan (1558–1603)

The Elizabethan period marked the culmination of the Tudor style. The main decorative features were rich, deep carving, bulbous turning, and inlays. It was a period of elaborate display in dress, manners, furniture and architecture. Elizabethan was a style for show rather than comfort (Figs 7.55, 7.56, 7.57).

Chairs and tables followed previous Tudor styles, with more elaborate decorations. Chests showed much improvement. The top lid gave way to doors on the front, and more drawer space turned the piece into a buffet or cabinet.

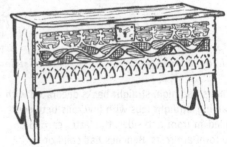

Figure 7.54 Early sixteenth-century planked chest with Gothic carved front; front and back are simply nailed to sides

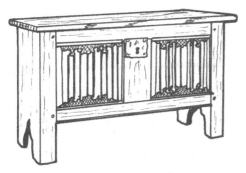

Figure 7.55 Late sixteenth-century framed chest with linenfold panelling; panels are held in grooves of frames to allow for 'working' of wood

Figure 7.56 Late sixteenth-century Elizabethan box stool

Jacobean (1603–1688)

Embracing the end of the age of oak and the beginning of the age of walnut, the Jacobean period was simpler in style than the Elizabethan. Ornament was less obtrusive but rather monotonously repetitious, straight line predominating in geometrical patterns of moulding, half-turning and diamond shapes. Turnings, plain or carved, were less bulbous. Upholstery became richer and carving shallower, but both retained Renaissance motifs.

During the Commonwealth under Cromwell, styles were simpler with a bare minimum of ornament, but the return of Charles II from France in 1660 brought about another design change. With the introduction of walnut, a more suitable timber for delicate work, Restoration furniture followed the luxurious styles of the Continent, influenced by French, Flemish and Italian ideas. Legs were more varied in shape and veneering, inlay, marquetry,

and silver and gilt ornament were introduced. Metal fittings appeared, such as handles, knobs and escutcheons, though they were more restrained in design than those of the Continent. Specialisation began and the work of the furniture-maker became separate from that of the carpenter. It is significant that the name 'cabinet-maker' now appears, reflecting the modifications in the design of the chest. With doors and drawers elaborately inlaid and decorated, this became in fact a cabinet.

Chairs were more comfortable with cushions and upholstery but still had straight backs, often with carved rails. Legs were turned in 'barley sugar', or twist, pattern with square sections for joints, and feet were in the form of knobs or buns. Farthingale chairs, without arms, were specially designed for the use of ladies wearing farthingale skirts.

Tables had heavy tops with moulded edges, turned legs, and rails near the floor on all sides. Drawers with fronts decorated with geometrical mouldings were fitted to side tables, whose shallow-carved front rails were supported by shaped valances or corner brackets.

Chests now took the form of cabinets and buffets with drawers and doors, although chests with lids were still made. Chests with drawers but no cupboards were introduced. Ornament was the same as for other furniture.

William and Mary (1688–1702)

With the arrival in England of the Dutch Prince William of Orange came new ideas in decoration, styles and types of furniture and also many foreign craftspeople. Walnut was still the predominant timber but veneering was used to a much greater extent, in many cases only the legs and main rails being solid. Edges and corners were protected with lipping and cock moulds. Turned legs of more varied design, with trumpet shapes and inverted cups a common feature, terminated in ball or bun feet. Scroll- and serpentine-shaped rails, flat rather than on edge, were fixed just above the floor. Rails and underbracing were pierced and carved. This was a transitional period between the straight, square frames with turned legs and posts of the Jacobean period and the graceful curved styles with plain and carved legs of the Queen Anne and Georgian periods.

Chairs, lighter than Jacobean, had high-raking carved and shaped backs, with woven rattan cane panels and seats. Legs were elaborately turned in scroll and cabriole shapes with horn feet and carved scroll tops. Upholstery was used, but springs had not yet been introduced.

Tables do not appear to have been made in large sizes during this period, although folding tables, dressing tables and writing tables, many with drawers, were made in increasing numbers. Construction and decoration were similar to those of other furniture of the period.

Chests of the older lidded type had virtually disappeared, being replaced by the more useful and convenient cabinet and bureau. The new social custom of tea drinking brought with it the introduction of cabinets for displaying china tea-sets. These cabinets had glass doors and pediment tops and often cupboards or drawers below. In some, the

bottom half was like a table and the top half a cabinet set back from the edge. The bureau or writing desk was a cabinet with drawers, either exposed or covered with doors. Legs were turned or square-tapered with serpentine rails similar to other furniture. Elaborate marquetry and matched veneers were common features, and drawer handles were of metal.

Queen Anne (1702–1714)

With walnut still the chief timber, graceful curves were now introduced into design (Fig. 7.57). Cabriole legs ended in pad, ball, and claw and ball feet, without bottom rails. Bracket feet were also used. Delicate carving in acanthus, husk and shell forms was a feature. Marquetry was less favoured and the beauty of timber grain was exploited by the wider use of veneers.

Chairs were lighter and more comfortable with few straight lines and featured lower, curved backs with a centre splat. Richly embroidered cloth coverings appeared on easy chairs.

Tables, made in the same wide variety as those of the William and Mary period, featured shaped aprons below the rails. Chests similar to those developed during the preceding era replaced

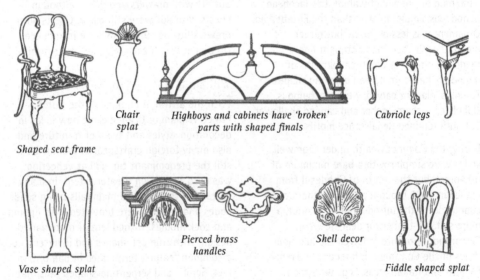

Chair splat

Highboys and cabinets have 'broken' parts with shaped finals

Cabriole legs

Shaped seat frame

Vase shaped splat

Pierced brass handles

Shell decor

Fiddle shaped splat

Figure 7.57 Characteristics of Queen Anne style

ornate marquetry with matched veneers for decorative effect. Writing desks were made with drop fronts, compartments for stationery and drawers fitted with brass handles.

Georgian (1714–1830)

The Georgian period includes the reign of the four Georges and the Regency period (1811–1820) when the Prince Regent, who was later to become George IV, ruled. It is one of the most important periods in the history of furniture. The age of walnut ended and was replaced by the age of mahogany and, later, the age of satinwood. With the end of the walnut age, the age of the cabinet-maker gave way to the age of the designer, which has been called the 'Golden Age of English Furniture'.

Early Georgian (1714–1745)

The change from walnut to mahogany required a change in construction methods. The early mahogany had less figure than walnut and to compensate for its rather plain grain, decoration in the form of panelling, moulding and carving again became prominent. Style followed that of the Queen Anne period, and it was not until mahogany became more popular that furniture became heavier and richly decorated. French rococo and baroque styles were the basis of much of the ornament and carving was predominately in the forms of lion and eagle heads, shells, acanthus leaves, claw and ball feet, ribbons, scrolls, swathes and clusters of fruit.

The designer associated with this period is William Kent, an architect who designed the furniture for his buildings. Kent's furniture had an elaborate 'architectural' style.

Chairs followed the Queen Anne style, but had more carving, lower backs with more ornately shaped and carved splats, and thicker cabriole legs with claw and ball feet.

Tables in large sizes seem to have been rare. Marble tops were often used for side tables and many small tables were made of veneered walnut, since the rather plain mahogany of

the time was unsuitable for veneers, the primary purpose of which is to show figure. Ornament was similar to that of chairs.

Chests followed the Queen Anne types but there were changes in structural detail. Panelling was used in frames, particularly for doors. The panels were often raised and carved, with moulding round the edge of the frame or applied to the panel in curved outlines. Presses came into more general use. They were deep, often with oaken sliding trays in the top section and doors with rule joints to allow the trays to be slid out. Tops were straight or pedimented and handles were of brass.

The age of the designers (1745–1880)

Thomas Chippendale (whose work covered the period 1745 to 1779), the Adam brothers (1760 to 1792), George Hepplewhite (1760 to 1790) and Thomas Sheraton (1790 to 1800) were not the only designers of the time, but rather the style leaders whose published works were used by many cabinet-makers. It is interesting to note that both Hepplewhite and Sheraton considered Chippendale's designs to be without merit, and their own were so much alike that it is often difficult to distinguish one from the other.

Thomas Chippendale's book, *The Gentleman's and Cabinet-maker's Directory*, published in 1753, was virtually a catalogue of designs. A practising cabinet-maker, he was able to adapt to his own style Chinese, Dutch, French and Gothic designs, generally adding rich decoration. His work is noted for the style of the legs, which were never plain, even straight legs being moulded. Carving and fretwork, never inlays, were used for decoration. Carving designs included acanthus leaves, fruit, ribbons, birds, flowers, claw and ball feet, lattice fretwork and Gothic tracery. Chippendale was a master in the use of mahogany, although he also used rosewood, beech and pine for carving, the latter usually japanned or gilded. Veneered work was in panels rather than overall, with construction of joints showing (Fig. 7.58).

Figure 7.58 Typical Chippendale style features

Chairs were notable for their good design and it is said that Chippendale developed chair-making to its peak. They were well made, strong yet graceful in line, although heavier looking than those of the other three designers. They were also made in a wide variety of designs. Those with cabriole legs had no stretcher rails while other types usually had two side rails with an intermediate cross rail, and sometimes a back rail. Backs were curved backwards and laterally, with elaborate shaped, carved and pierced slats and horizontal rails. Arms were curved in plan, elevation and section. Chinese designs, sometimes with turned front legs, had vertical members between arms and side rails and carved apron pieces and brackets under the front rails.

Tables were made with the same range of leg designs as chairs, but the top carving was carried farther down on the knee and also on the inside edges. A new type of leg was introduced, built up in a cluster of turned pieces. Among the wide variety of tables, tea-tables and side tables were popular, the latter having carved front rails, and tops with carved, moulded edges. Tripod tables had three shaped and carved legs radiating from a centre pillar, and the top often had a raised, carved and scalloped edge, known as a 'pie crust' edge.

Chests had long been superseded by more elaborate and useful cabinets, buffets, bookcases and writing desks. Doors showed a pattern of narrow bars and glass panels or veneered panels with shaped and carved mouldings planted on, and the edges of the frame were also moulded. Drawers and fall fronts were fitted to writing desks and bookcases. Serpentine shapes with much detailed carving, and scroll or carved legs were used on tables and commodes. Brass handles and escutcheons decorated doors and drawers.

Robert Adam is the best known of the four Adam brothers, who were all architects and published a number of books on architecture. Though they were not cabinet-makers and their books did not deal specifically with furniture, their influence on furniture design was profound.

Like William Kent in the time of George I, Robert Adam designed not only the building but the furniture and furnishings to go with it.

He was a great traveller and studied the Classic styles of architecture in many parts of the world, so that his own style was decidedly Classic. His work is delicate and straight-lined in contrast with the preceding rococo curves, and features inlay, light-relief carving in the form of Greek key patterns, vases, husks, festoons, plaques, fans, urns, discs and ellipses. Painted designs by prominent artists were also incorporated. The ornament was always carefully designed so as not to overload and spoil the overall graceful style (Fig. 7.59). Both Chippendale and Hepplewhite made furniture to Adam designs.

Chairs were much lighter than those of Chippendale, with more curves than other Adam furniture, whose typical outlines were straight and rectangular. Legs were straight or lightly curved or tapered, turned or square and decorated with shallow carvings, applied carvings and gesso work, inlay and mouldings in the form of flutes and reeds. Gilt and silver were often used on the ornament. Backs were made in many shapes, including shield, lyre and goblet. Centre slats were shaped and carved and upholstery was designed to match the other furnishings.

Tables included dining tables with a flap top and a pivoted fifth leg for support. A drawer right across the front was another innovation. Side tables were made with curved rails between the legs. Adam designed the forerunner of the sideboard, a composite piece consisting of a side table with a separate, taller pedestal cupboard at each end.

Chests were interpreted as tall wardrobes, chiffoniers and cabinets, painted and gilded, with straight tops, plinth bases and brass handles.

Table 7.1 *Summary of furniture styles*

Year	1500			1600			
Reigning monarch	Henry VIII 1547	Edward VI 1552	Mary I 1558	Elizabeth I 1603	James I 1625	Charles I 1649	Cromwell 1658
Age							Carpenter 1660
Wood used							Oak 1660
Periods	Gothic influence		Tudor	1603		Jacobean 1660	Commonwealth 1659
Tables	Trestle, refectory	Side	Draw-leaf extension		Gate leg		
Chairs	Trestle, stools, forms; very few chairs	Settles	More chairs used	Farthingale		Upholstered	
Chests	Carved from solid logs; some panelled	Cupboards, buffet, settle developed from chest	Full-panelled; door on front	Drawers		Chest of drawers, china cabinets; glass doors	
Legs	Square		Bulbous turned	Urn and bobbin turned	Twist turned		
Decoration	Gothic, linenfold, romayne; carved chamfered, incised		Deep carving in Renaissance designs; acanthus leaf, floral and geometrical patterns	Applied moulding in geometrical patterns; split turnings and lozenge applied; carved bulbous turning; veneering		Bracket feet; bobbin turning	
Important developments	Gothic designs based on masonry ornament	Wider use of furniture; Gothic motifs still used	Wide deep mouldings, plain and carved	Repetitive designs in carving; greater use of turning	Backs of chairs inclined	Great destruction of churches during Cromwell's time	

		1700				1800
Charles II 1685	James II	William and Mary 1702	Anne 1714	George I 1727	George II 1760	George III 1820
				Cabinet-maker 1750		Designer 1806
		Walnut 1725			Mahogany	Satinwood
Transitional 1668		William and Mary 1702	Queen Anne 1714		Chippendale 1780 Hepplewhite 1790	Adam 1792 Sheraton 1806 Georgian 1830
	Writing desks, card tables			Veneered oak still used; drawers added		Increasing number of types and uses of tables
	Straight back, thin leather covers, rattan seat and back		Curved back with centre slats; some chairs with stretchers	Ribbon back, pierced slats, back sloped	Shield back, oval back	Low, square back
			Framed base; drawers hidden behind doors	Writing desks, bureaux		Pediment tops, rich ornament
	Scroll inverted cup; trumpet shape; part of underframe		Cabriole	Cabriole and straight; claw and ball foot, hoof and club foot; fretted	Straight turned	Straight turned
	Turned trumpet, serpentine frames; scroll-shaped legs; carved; elaborate marquetry			Rich tapestry; pierced and carved chair backs; floral and acanthus leaf carving		Strip inlays, veneers, delicate moulding, painting and gilding
	Transitional period; beginning of cabinet-maker; change to walnut required specialist skills		Influence of Dutch designers and draftspeople; restrained ornament during Queen Anne period			Georgian decoration lighter than that in William and Mary period; Chippendale chairs richly carved; publication of books on furniture design

George Hepplewhite was a cabinet-maker, though few pieces of his work still exist and most of what we know of his designs comes from his book, *The Cabinet-maker's and Upholsterer's Guide*, published two years after his death. Influenced by Adam and the French styles, his designs were simpler, less ornate and lighter in construction than Chippendale's and closely resemble those of Sheraton (Fig. 7.60). Restrained decoration included inlay; light carving in the form of husks, flowers and Prince of Wales feathers; painting and gilding; and painted Classic designs. Legs were mainly square or turned tapered or in a French cabriole style.

Chairs were outstanding in design. Hepplewhite's shield back is a particularly good example of its type, and he designed also hoop, oval and heart shapes. Rails were used between all legs except cabriole.

Tables of many types were made. The Pembroke table with hinged side flaps and hinged supporting brackets, semicircular side tables, and card tables with fold-over tops were all fashionable.

Chests in the form of cabinets, buffets, tallboys and wardrobes were all in typical Hepplewhite styles. The design in which a centre table and pedestal cabinet were combined to make a sideboard was probably devised in collaboration with Thomas Sheraton.

Thomas Sheraton probably never had his own workshop or made any furniture, but his designs are known from his books. The first, *The Cabinet-maker's and Upholsterer's Drawing Book*, published between 1791 and 1794, not only showed designs but included instructions in geometry and perspective drawing, as well as many ingenious ideas on mechanical movements and fittings for furniture.

Satinwood had now become popular and Sheraton designed for its use in construction, veneers and inlays, following the pattern of the lighter styles of Adam and Hepplewhite. Delicate ornament in the form of strip inlay, marquetry and some low-relief carving was in Classic style, with circular and oval shapes, urns, acanthus leaves, scrolls, fans, lyres, festoons and bows predominating. Paint, japanning and gilding were frequently used, overall or for panels and motifs (Fig. 7.61).

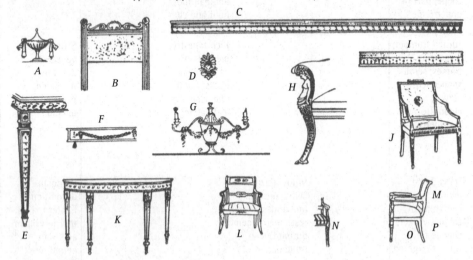

Figure 7.59 Adam brothers designs, 1760–1782: (A) classic urn; (B) solid panel back; (F,C,I) daintily carved mouldings; (D&G) floral swags and pendants; (E,N&O) slim, tapered round or square legs; (H) human figure; (J) upholstered seat and back; (K) decorative moulding; (L) arms supported by extension of legs

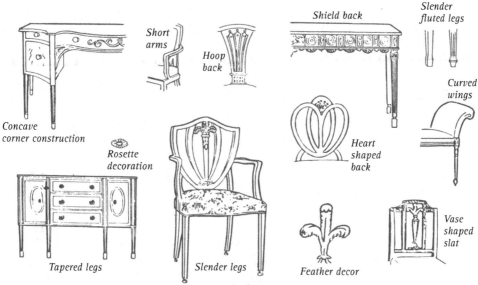

Figure 7.60 Features of Hepplewhite style, 1760–1790

Sheraton's later designs, in Empire style, introduced as decoration all kinds of naval emblems, such as ropes, anchors and lifeboats, but they lacked the elegance and grace of his earlier work and are generally considered inferior.

Chairs by Sheraton, like those of the three other style leaders, were particularly well designed. He broke away from the curved outline for chair backs, making them lower and squarish, with vertical splats turned or carved. The top rail often had a panel or tablet in the centre or a festoon carving joining the slats. Overall, the chairs had a square, straight lined appearance, and decoration consisted mainly of fine fluted or reeded mouldings, strip inlays or light carving, and paint and gilding. Legs were tapered, square or turned, never cabriole, and seldom had rails between. Arms were designed to come from the top of the back in a continuous sweep and were supported by turned uprights from the side rails.

Tables were light and graceful with tops veneered or faced with marble. Legs were similar to those of chairs and seldom accompanied by stretcher rails. Serpentine fronts were common and most tables were fitted with drawers, inlaid and with lip mould or cock beads at the edge.

Chests are represented by sideboards, with serpentine or round ends on the front, marble tops and tubular or turned brass rails along the back. Drawers and cupboards were concealed behind sliding roll doors. Veneered and inlaid writing desks, bureaux and cabinets with elaborately fitted cupboards often had ingenious mechanical opening and closing devices.

THE NINETEENTH CENTURY

With the nineteenth century began an era of great changes in all design brought about by the sweeping political and sociological changes of the time. Late Georgian and early Victorian England, in common with the rest of Europe during the first half of the century, felt the impact of the Napoleonic Wars and the rise of the French Empire. It was also influenced by the new interest in Classic styles stimulated by archaeological discoveries in Italy, Greece and Egypt. The second half-century was marked by the social development attendant on the increased use of machines, which began to

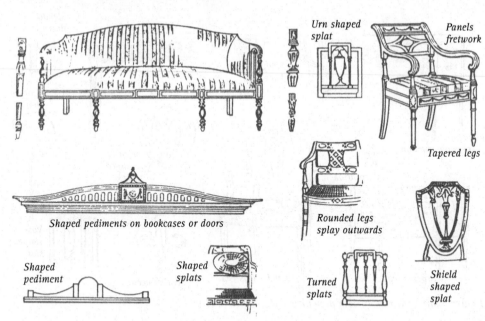

Urn shaped splat

Panels fretwork

Tapered legs

Rounded legs splay outwards

Shaped pediments on bookcases or doors

Shield shaped splat

Shaped pediment

Shaped splats

Turned splats

Figure 7.61 Distinctive Sheraton patterns, 1780–1806

supply mass-produced furniture as well as many other commodities to meet the demands of a growing affluent middle class.

Regency (1811–1820) and Early Victorian (1820–1850)

Regency furniture was greatly influenced by the French Empire style and the neoclassicism stemming from 'finds' among the antiquities of Greece, Italy and Egypt. Lacking the graceful lines and lightness of the best eighteenth-century furniture, Regency ornament and decoration followed these ancient styles, often combining a curious mixture of all three in the one piece.

Designers associated with the Early Victorian period were Thomas Hope and Sir John Soane. Design became stiff and formal, characterised by heavy carving, brass mounts and inlay, and over-ornamentation with scrolls and reed mouldings, many of them machine-made.

Chairs with outward-curving 'sabre' legs had upholstered backs shaped in S-bends matching the ends of sofas. Cheaper furniture was made with straight lines. Tables were made in a great variety of shapes, some with lyre-shaped ends in place of legs. Chests, in rosewood and other dark-coloured timbers, were produced as cabinets, desks, bureaux, sideboards and similar pieces.

Late Victorian (1850–1900)

The Late Victorian era was the true beginning of the machine age, where the individuality of the craftsperson that had formerly marked the design and decoration of handmade pieces was lost in the welter of applied ornament in the form of machine-made scrolls and mouldings and novelties of all kinds. In the transition, many new techniques had to be acquired, mostly by trial and error, to utilise the new machinery to maximum productive and economic advantage.

Chairs were made in a wide range of styles, including the popular 'balloon back', and were padded with horsehair or coiled metal springs. Comfort was considered important and easy-chairs became fashionable. Bentwood furniture was developed in Austria by Thane and 'austrian chairs' became a generic term.

Tables for Victorian dining rooms were often circular with tripod legs, and extension tables with a screw mechanism were also popular.

Chests reached their peak in the sideboard, made up to seven feet long in mahogany or wood stained to imitate it. A lower cabinet contained drawers, cupboards and a multi-shelved superstructure on which crockery, glassware and knick-knacks were displayed round a central mirror.

Toward the end of the period, the designers William Morris and Ernest Gimson, dissatisfied with the prevailing styles, attempted to improve them but without much success.

CONTEMPORARY FURNITURE

In common with design in other fields since the turn of the century, furniture design has seen sweeping changes which continue today. These are worldwide, so that the coming styles in Australia can reasonably be forecast from trends observed overseas, especially in Europe.

A characteristic of the new age is that new types of furniture pieces are constantly being introduced while some older ones are being discarded. For example, the advent of television created a demand for lighter and more mobile easy-chairs and greater numbers of small, low tables. The trend towards outdoor living calls for lightweight, comfortable, weather-resisting chairs, lounges and tables. The need for space-saving, built-in cupboards and fitments and adaptable unit furniture, some of which comes in the form of mass-produced 'knock-down' parts for assembly on site, has naturally influenced design. But probably the greatest single factor is the mass production process itself, where highly complicated and automatic machinery demands styles that can be produced in quantity and allow for variety to be achieved

through interchangeable parts, a choice of timber finishes, and a range of types, colours and patterns in upholstery coverings.

Figure 7.62 Typical furniture using European beech veneered particle board. Entertainment Module combining storage and display by Covemore Designs

Certain fashion trends in styles, materials and finishes may be observed in current patterns in domestic furniture.

Bedroom furniture, apart from some pieces in simplified forms of traditional or period designs, is rectangular in shape rather than curved. It features built-in wardrobes or those composed of units capable of vertical or lateral extension. Most headboards nowadays are higher, with some form of decoration such as carving, turning, lead lighting or wrought iron. Bedside cabinets have replaced extended side cupboards. Legs, where they are used, are short and turned. Handles are either contoured to blend with the design or styled to provide contrast or add colour. Wood grain with gloss lacquer or polyester finish forms the main decoration, though paint in light shades with coloured trim is also popular.

Dining-room and lounge-room furniture is often made in matching pieces in keeping with the present trend to combine both rooms in one (Fig. 7.63), but different designs may be blended

(Fig. 7.64). Low, wide, rectangular buffets with sliding or hinged doors contain drawers and cocktail cabinet and may be composed of unit pieces. Legs, straight or turned, are often part of a base frame. The round or the elliptical (sometimes miscalled oval) shape is favoured for dining tables, many of which are provided with an automatic extension apparatus. Turned legs support the top, which usually has a plain or

Figure 7.63 Modern contemporary upholstered lounge and furniture follows clean simple lines

Figure 7.64 Combined lounge and dining interior, showing mixed furniture styles

Figure 7.65 Contemporary dining-room setting, showing pleasing balance of design elements

simple moulded edge. Occasional tables, mostly rectangular, may have wooden or metal legs and frame surmounted by a top in natural timber, plastic laminate, marble or ceramic. Some have

Figure 7.66 Fabric covered lounge. Note square corner table with table lamp and wall picture that blend into the decor

Figure 7.67 Contemporary lounge, and corner occasional table arms and back and turned or shaped legs. Occasional chairs often have detachable cushion seats and backs, and lounge chairs may be units arranged in sets or separate pieces fully upholstered and with adjustable backrest and footrest

Figure 7.68 Lounge and dining furniture designed to blend into a modern setting. Note leather covered lounge

a shelf of wood or framed woven cane. Sets of small tables are made in graduated sizes and without rails or shelf so that they nest together when not in use, thus saving floor space. Chairs are light in design, with shaped arms and back and turned or shaped legs.

Kitchen furniture is mainly built in, with sink, stove and refrigerator incorporated in units, the stove often accompanied by a separate wall oven. Tables and chairs with metal frames and legs, the chairs upholstered in vinyl and the tables and workbenches topped with plastic laminate are still popular although solid timber furniture is gaining in popularity. Cupboard and drawer handles and edging to tables and benches are decorative as well as utilitarian. Woodwork is clear-finished to show the grain, or painted, or covered with melamine in some form.

Figure 7.69 Contemporary kitchen combines the warmth of wood with stainless steel and the brightness of laminated plastic-coated particle board.

Outdoor furniture has revived the patterns of cast-iron filigree work and now executes them in cast aluminium, or even plastic. Wrought iron and tubular aluminium are also used, as well as natural-finished or painted timber. Chair seats and backs are of non-sag plastic webbing in fast colours and cushions are also covered with plastic fabric.

New materials and methods

As in building, new materials and methods have made possible the development of many new ideas in furniture design and construction.

Particle board may be used for carcass construction and shelves without the need for framing. Being almost inert, it is particularly suitable as a base for veneer and plastic laminates. An extensive range of sizes is available. Coreboard, plywood and hardboard, though not among the newest materials, now offer a wider range of finishes, sizes and qualities, thus increasing the scope of their application.

Plastic laminate in a variety of types and finishes is finding a growing number of uses. Its durability and resistance to heat and acids make it an invaluable surfacing material (Fig. 7.69). Veneers of timber or plastic laminate are applied to particle board or coreboard for use in carcass and shelf construction, in which these made-up boards have largely replaced solid wood.

Vinyl, used extensively for surfacing and coverings, is durable, resistant to heat and most stains, easily cleaned and comes in a wide range of thicknesses, colours, textures and patterns.

Synthetic fabrics, woven cane, and other materials such as thin cork are used for upholstery or to introduce contrasting texture. Plastic webbing for chair seats and backs is available in stretch or non-stretch types and in many finishes and fast colours.

Springs and strip rubber form the bases for upholstery and provide easy adjustment for overlaid cushions. Plastic and rubber, granulated or as sheets of foam or sponge, make filling for cushions and padding for chair backs. Sheet thicknesses range from 2 to 150 mm. Foam and sponge are also available formed to shape for particular applications, such as seat padding for dining chairs.

Handles, fasteners and hinges reveal many new developments. Handles of various shapes are produced in wood, metal and plastic. Fasteners are of magnetic, snap, ball, roller and touch types. Hinges have shown great improvements, notably the introduction of concealed and semi-concealed kinds. Hinges of nylon and drawer slides of nylon and plastic operate quietly and smoothly with very little wear. Shelf supports and drop mechanisms are varied and efficient.

Edgings, inserts, mouldings, and leg-end caps in wood, metal, plastic, or in combinations of these, are used to cover raw edges as well as to decorate.

Legs of metal, wood or plastic, usually round and tapered but also square-tapered, may form part of a base frame or may be screwed to metal supports attached to the base.

Finishes, unlike the relatively fragile French polish, which is affected by heat, are now mostly of the two-pot type, highly resistant to heat and acid.

Adhesives to meet every requirement are now available for application by brush, roller, scraper or spray. Allied to adhesives are the many kinds of cramps and presses, some of the more elaborate presses operating in conjunction with radio frequency glue-curing.

ARCHITECTURE AND BUILDING IN AUSTRALIA

The story of Australian architecture is part of our history. Closely linked with all phases of the nation's development, it began with the arrival of the First Fleet in 1788 and is still being written in the buildings of today.

Although it is not possible to set exact dates for the particular architectural styles, for they all overlap in time, it is convenient for study purposes to regard them as falling within certain flexible time limits. Thus it is generally accepted that the years between 1840 and 1850 saw the end of the early colonial and Georgian periods to be followed, after the gold rushes of the 1850s, by the Gothic and Classic revivals which marked the Victorian era, that is, the period coinciding with the reign of Queen Victoria. Art Nouveau (literally, 'new art') makes a brief transition period from the late nineteenth century to the early twentieth until the beginnings of what we may call 'twentieth-century contemporary' styles.

EARLY ARCHITECTURAL STYLES

New South Wales and Tasmania must be our sources of information regarding the earliest buildings, for it was in these states that colonisation began—at Sydney Cove in 1788 and in Tasmania, then called 'Van Diemen's Land', in 1803. The story begins later in the other states, for they were not settled by Europeans till the years between 1825 and 1836 and they lagged in population and development until the discovery of gold in the 1850s brought an influx of immigrants and wealth.

Earliest buildings

The First Fleeters brought with them to Sydney a few tents and, for Governor Phillip, a prefabricated house, which was virtually no more than an improved kind of tent. So for the majority, who, it must be remembered, knew nothing of the natural resources of their new country, shelter was an urgent need.

The first buildings were wooden framed and clad with split trunks from the cabbage tree palms, which were plentiful, or with slabs split or pit-sawn from other native trees and sometimes driven into the ground for greater stability. Roof and walls were covered with clay, which, however, was soon softened and washed out by rain. Clay strengthened with wattle twigs was also used for cladding and this wattle and daub construction was made more satisfactory by the addition of tamped earth. Another building material, called pisé, consisted of a mixture of clay, earth and gravel rammed between boards, which were removed when the mixture hardened. Examples may still be seen in some country areas today. Roof coverings were sometimes palm fronds, sheets of bark or a thatch of rushes, most of the latter being obtained from the area still called Rushcutters Bay. Later, she-oak shingles were used and tiles made from puddled clay, while corrugated iron was being imported before 1830. Construction methods had to be simple. Most fastening was done with wooden pegs since nails, which were at first handmade and therefore scarce and expensive, did not come into general use until about 1850.

Despite an acute shortage of building materials and hardware, suitable tools, and skilled tradespeople, and in the absence of any person qualified by skill or experience to organise and control a building team, work was begun in May 1788 on Australia's first building of any architectural pretensions, the first Government House, in Bridge Street, Sydney. Foundations were of stone, of which ample supplies of good quality had been discovered; walls were of bricks made from clay found within a mile or two of the settlement, and were finished with stone quoins, or corners; and the roof was of burnt clay shingles. Lime for mortar was burnt from shells collected from the foreshores, for it was not until forty years later that limestone deposits were found, and these well away from the coast. The colony's first staircase was installed in this building.

For the next twenty years, building scarcely kept pace with the new colony's expansion. When Governor Macquarie arrived in 1810, he was confronted with the problem of increasing numbers of convicts, for whom work had to be found. So he embarked, in spite of a marked lack of co-operation by the colonial office in England, on his enormous building program, in which the architect Francis Greenway was to figure so prominently. The full list of works constructed in New South Wales and Tasmania during his twelve years of office includes 166 buildings of brick, stone, or weatherboard; 4 bridges; 7 harbour works; many log buildings, dairies, timber yards and markets; repairs to many existing buildings; and more than 500 km of roads.

The prevailing style of architecture was Georgian modified to suit the Australian environment. Most buildings were single-storeyed, with a few of two storeys.

Characteristics of Georgian architecture

Relying on good proportions, sensitive use of materials and careful detailing rather than applied ornament, the Georgian style is characterised by its symmetry. Wings and chimneys balanced each other on either side of the centre. In domestic architecture there must be three or five openings on the front facade; thus if the centre opening were a door, it must be balanced on each side with either one or two windows.

Glass areas, including fanlights, were broken up by bars into small panes, since glass could not then be produced in large sheets, and six- and twelve-light sashes were common (Fig. 7.70). Shutters to doors and windows were protective as well as decorative (Fig. 7.71).

Columns of wood or stone supporting verandas and porches were often classical in form, usually of the Doric Order. Walls were natural brick or stone. Later in the period they were stuccoed, or plastered over, with mouldings run in cement or plaster instead of stone, and buildings finished in this way are said to be in 'Regency' style.

Figure 7.70 Lancer Barracks, Parramatta, NSW; built 1820; John Watts, architect; Georgian style typified by symmetry of openings and chimneys, simple straight lines and lack of ornament; walls of brick showing signs of weathering; semicircular fanlight over entrance door, small panes in windows; original building had no veranda

Figure 7.71 Parliament House, Sydney, originally part of the Rum Hospital, built 1817; architect unknown; Georgian balance of doors, windows, and steps; stone walls; turned columns with moulded bases and caps; semi-circular fanlights to doors, with arches projecting beyond face of walls; emphasis on horizontal lines

Characteristics of Gothic Revival

Copying the pointed style of medieval churches and the crenellated battlements of castles, Gothic Revival in Australia is seen in both church and domestic architecture. Long, pointed windows, steeply pitched roofs, tall, sharp steeples and spires, and richly carved ornament are characteristic features (Figs 7.72, 7.73, 7.74). Buildings of all kinds showed

Gothic influence, a notable example being the Government House stables in Sydney, originally designed by Greenway in 1817 and later converted and still used as the Conservatorium of Music.

Characteristics of Classic Revival

The new affluence brought by the discovery of gold found an outlet in impressive buildings. Numbers of large public buildings, such as town

Figure 7.72 St Marys Cathedral, Sydney; built 1865; William Wardell, architect; typical Gothic style, with accent on vertical lines, tracery windows and decorated stonework; plain buttresses

Figure 7.73 St Andrews Cathedral, Sydney; commenced 1837 by James Hume, completed 1846–1847 by Edmund Blacket; another example of Gothic style; compare design of tracery windows, decorated gables and moulded buttresses with those of St Marys Cathedral

Figure 7.74 Bishopscourt, Darling Point, Sydney, residence of Anglican archbishop of Sydney; built 1846–1853; Edmund Blacket, architect; Gothic style; stone walls, steeply pitched roof, dormer windows; projecting fireplace an external feature; oblique chimneys; flat pointed arch, deeply moulded with carved corbels over entrance, and three long windows above with stepped moulding; wooden barges, with moulded finials

Figure 7.75 Courthouse, Berrima, NSW; built 1836–1838; Mortimer Lewis, architect; used as courthouse until 1900; Greek Revival style; Doric columns with flat half-columns or pilasters on side wall and corner; simple style without ornament, relying on good proportions for beauty; plain architrave and pediment

halls, cathedrals, hospitals and office blocks were erected during the Victorian era. Something ornate yet classical seemed to be called for, so the ancient Greek and Roman styles were invoked and given mouldings, columns and ornament more elaborate than the originals. Doric, and more often Corinthian, columns stood free or were built into the walls to show as half-columns, and were surmounted by the usual entablature and pediment (Fig. 7.75). In houses, columns appeared in conjunction with arches in hallways and between rooms, and windows and doorways were similarly enriched. It is interesting that churches and cathedrals retained the Gothic style.

Decoration in the Victorian era

Rich ornament was the prevailing fashion from the 1840s to the 1880s, the prime years of the Victorian era. Architects did not attempt to form any new style but turned, as we have seen, to the traditional Gothic and Classic, often combining them and adding more and

more ornament to make them look different. Architectural beauty was measured by the amount of decoration that could be crammed on to a surface, and the facades of buildings were crowded with columns, figures, floral swathes, carving, parapets or pediments capped with urns, and roofs with turrets or domes (Figs 7.76, 7.77). This was the era of applied ornament in the form of cast iron, stucco and mouldings run in plaster and cement. Internally, skirtings, architraves, picture rails, ceilings and cornices were all decorated with thick, wide mouldings in intricate floral and geometrical patterns, and with reeds, grooves, sinkings, ovolos, ogees and scotias. Ceilings were lofty, above high, narrow windows with small, dark glazed panes. Doors were topped with fanlights and walls were covered with paper in floral patterns. Colours were invariably dark, deep-stone and chocolate being favourites. Woodwork was grained, marbled or gilded—nothing could appear in its natural state, everything had to imitate some other material.

Not all architects agreed with the florid concepts of the boom period. One notable

Figure 7.76 Colonial Treasurer's Building, Sydney; built 1878; James Barnet, architect; now Chief Secretary's Department; accent on horizontal lines gives squat appearance; statues in niches on corner; square dome; alternating triangular and curved pediments to top most windows; stone balustrade with urns; Doric and Ionic columns and pilasters

Figure 7.77 Courthouse and Post Office, Bathurst, NSW; built 1878; Greek Revival style of balanced design; double and single composite columns at entrance; colonnade along front; semicircular windows in octagonal dome; cupolas above entrance and dome; stone pavement

dissenter was Sir John Sulman, who crusaded for town planning and more straightforward building design. Sulman, whose name is perpetuated by annual awards for painting and architecture, was considered very advanced, yet his own work shows only slight modification of the prevailing Victorian ideas. These persevered, becoming more and more elaborate until a series of industrial troubles in 1893 brought bankruptcy and financial ruin to many individuals and business firms. Building practically ceased for about two years, to begin again with a new concept of design expressed in Art Nouveau.

Characteristics of Art Nouveau

Art Nouveau probably began in Belgium and England in the 1880s. It was featured at the Dresden Exposition in 1897, and spread throughout the world. Based on the theory that 'nature abhors a straight line', it broke with all traditional and classical styles, using flowing curves based on floral designs. Though it was in vogue for the relatively short period of ten years or so, reaching its zenith about 1902, there was hardly any type of building or any article of furniture that was not touched by the sinuosities of the 'new art'.

In Australia it found expression in surface decoration rather than as a factor in overall building design, and motifs of Australian native flowers, leaves and gumnuts sometimes replaced the traditionally European roses, tulips and vines.

THE FIRST ARCHITECTS

Many houses, hospitals, barracks, stores, gaols, courthouses and churches had already been built before the arrival in 1813 of Daniel Mathew, the first qualified architect in Australia.

Daniel Mathew, who came as a free settler to practise architecture, drew many plans for buildings and evidently received some private commissions, but no detailed record of his work between 1813 and 1823 is available. He was responsible for only one government building, the colonial secretary's house, a Regency-style structure in Bridge Street, Sydney, which stood until 1915. He waged a constant, but apparently unsuccessful, campaign for more work from the government, and eventually abandoned architecture to become a farmer and sawmiller.

John Watts, a lieutenant in the army, came to Sydney in 1814 as aide to Governor Macquarie and received no additional pay for his architectural work. Some examples still exist in Tasmania as well as in and around Sydney, including the oldest existing military establishment in Australia, now called the Lancer Barracks, at Parramatta (Fig. 7.70). Watts returned to England in 1819, and was reposted to Australia eighteen years later as postmaster general, but took no further part in architectural work.

Francis Greenway, convicted of forgery, was transported to New South Wales from England in 1814, and as Macquarie's protégé and a master of the Georgian style, was to exercise the most profound influence on early Australian building. He was first called by Macquarie in July 1814 regarding plans for a town hall and a courthouse. Granted a ticket-of-leave, in December, he was advertising his services as an architect and planner and evidently obtained some private commissions to provide him with a living, for all government work he did without pay. In 1816 he was appointed civil architect and assistant engineer to plan and erect government public works at a salary of three shillings and eight pence a day and held this post until 1822. His first assignment was a survey of the Rum Hospital, part of which is now Parliament House (Fig. 7.71). Many of his buildings are still standing, though none entirely in their original form. His three churches—St James in Sydney, St Lukes at Liverpool, and St Matthews at Windsor—are all accepted as fine architectural examples. Hyde Park Barracks has

been named as Greenway's masterpiece, though encroachments by other buildings now cloud the grand scale of the original plan with its 20 × 30 m compound. For this work he was granted a full pardon in 1817. Other important works by Greenway include the Macquarie Lighthouse (a replica now stands at South Head), Windsor Courthouse, the Female Factory of Parramatta, as well as a toll gate, a fort, a market house, a hospital, a police office, a quay and dockyard, and a project for town sewers and water supply. He was also engaged in preparing drafts for the building regulations that were a part of the organisational improvements introduced by Macquarie. All this was in addition to his private practice. In 1822, Greenway was dismissed after quarrelling with Macquarie and retired to the 800 acres of land granted to him at Raymond Terrace on the Hunter River and died there in 1837. It is difficult to ascribe definite dates to any of his work but there is no trace of any building designed by him after 1828 (Fig. 7.78).

There were other designers who were contemporary with Greenway, some of whom called themselves architects. James Smith

Figure 7.78 'Hobartville', Richmond, NSW; built 1828; may have been designed by Greenway; western elevation; a 2750 mm veranda flanks other three sides; eastern elevation is typically Georgian with straight front and entrance porch; walls of brick in English bond, with piers at corners; small panes in shuttered windows; cast-iron veranda posts

arrived as a free settler in 1815 and often opposed Greenway. Frank Lawless, a foreman-bricklayer of government gangs, designed convict barracks at Parramatta and possibly St Pauls church at Campbelltown and the Benevolent Asylum. Standish Harris replaced Greenway as government architect but, apart from a report on existing buildings, did very little except the wall surrounding the old gaol at Darlinghurst, now the East Sydney Technical College. In private practice he designed Scots Church, Sydney (now demolished). George Cookney succeeded Harris in 1824 but held office for little more than a year, during which time he designed the La Perouse monument near the entrance to Botany Bay.

Edmund Blacket was to the second half of the century what Greenway had been to the first. Blacket arrived from England in 1842 and became the leading exponent of Australian Gothic. In 1847 he was the inspector of Church of England schools and diocesan

Figure 7.79 Part of Sydney University (southern elevation); built 1854–1860; Edmund Blacket, architect; main east front reputed to be smaller copy of Westminster Hall, London; a fine example of Gothic Revival architecture; compare windows and ornament with those of St Marys and St Andrews cathedrals (Figs 7.72, 7.73)

architecture. In 1849 he was appointed colonial architect but resigned in 1853 to practise privately, concentrating on the building of Sydney University (Fig. 7.79), St Andrews Cathedral, Sydney (Fig. 7.73) and St Stephens church at Newtown, as well as a number of small churches. Before his death in 1883, many other designers had been inspired by his work to build houses in the Gothic style with steeply pitched roofs, gables instead of hips and the effect of carved stone tracery achieved with shaped and pierced wooden bargeboards and similar details. Whole terraces of houses as well as commercial buildings were designed with Gothic facades.

EARLY BUILDINGS

To appreciate the sweep of architectural development from the beginnings of colonisation to the end of the nineteenth century, it is convenient to group the buildings according to their purpose under the three headings of domestic, commercial and public buildings.

Domestic buildings

Following the slab or wattle and daub huts of the earliest settlers, colonial cottages from 1815 to 1840 were in the Georgian style. They were generally single-storeyed, with main rooms at the front divided by a hall and either one or two pairs of windows placed symmetrically at either side of the central entrance door. A separate kitchen and perhaps a storeroom were provided, but there was no bathroom or laundry. Roofs, of bark or shingles and later of corrugated iron, were hipped at each side of the building, and many of the early examples were without the verandas that later became a feature, especially in country areas. When verandas were included, they were generally of much lower pitch than the main roof and were stone-floored, running across the front of the building and sometimes along two sides as well. Rooms opened to the verandas with a door or french windows, usually shuttered (Fig. 7.80).

Figure 7.80 Experiment Farm Cottage, Parramatta, NSW; built 1798 by surgeon John Harris; site of James Ruse's Experiment Farm; during restoration by the National Trust it was found that ceiling joists had been dovetailed and dowelled and every piece of roof timber had been numbered with Roman numerals; joinery was of cedar and roof of oak shingles, since replaced with flat iron; bricks were sandstock with mortar made from shell lime and sandy loam, and flooring was of pit-sawn timber; building has Georgian balance; new tallowwood columns are of similar style to originals

Country homesteads were of much the same plan as the cottages (Fig. 7.82), but larger and almost invariably with a wide veranda on at least three sides (Fig 7.83). Veranda posts were usually much larger than strength demanded and were often of Classic design with turned base and cap. Kitchen, storeroom, servants' quarters, workshops and stables were separate, often built round a courtyard or driveway and sometimes linked by covered ways to the main house (Fig. 7.84). The colonial cottage and homestead reached their peak in the 1830s. It is interesting that inns and hospitals were similar in pattern to large homesteads.

Terraces of houses were built in the city and suburbs from the middle of the nineteenth century in order to house as many people as possible within walking distance of their work. With an average frontage of 4.5 metres these houses were often of three storeys, one at street level, one below and one above. Most had a front veranda opening to the street and a balcony above. Rooms were small, much space being taken up by staircases, and bathroom, toilet and laundry facilities were simply fitted into some corner of the basement, though washing was often done in the small backyard with water heated over an open fire.

Figure 7.81 Government House. The most sophisticated example of Gothic Revival building in NSW. The building was constructed between 1837 and 1845, and today features a magnificent collection of furnishings and decoration that reflects the changing styles of the 19th and 20th centuries. The building is located in the Domain, Sydney NSW.

Figure 7.82 Elizabeth Farm, Parramatta, NSW; built 1793 by John Macarthur; oldest farmhouse still in existence; little remains of original main wing, as it was altered five times in first seventy years; brick walls; hipped roof; original wooden veranda posts replaced with cast-iron supports

In contrast with the terraces, mansions were built in spacious grounds (Fig. 7.78), with large, stately rooms and ornate exteriors. Every conceivable shape of roof was used, in conjunction with gables, hips, valleys, turrets, domes and cupolas. Internal finish and decoration matched the elaborate exteriors.

Figure 7.83 'Burrundalla', Mudgee, NSW; a fine example of Georgian architecture; brick walls in English bond; doors with shutters opening on to wide, stone-floored veranda; good proportions, with accent on horizontal lines; no ornament

Figure 7.84 Covered way at 'Dabee', Rylstone, NSW; well-worn stone path; octagonal timber posts with moulded bases; six-light sashes; shutters; stone walls

In spite of all their elegance, such houses hardly provided easy living, since kitchen, storerooms and bathhouse were often separate from the main building, perhaps joined to it with a covered way.

Regency villas from the 1830s met the needs of people wealthy enough to want something more ostentatious than a cottage. They were often two-storeyed, in stone or cement-rendered brick, Georgian in style, but with bay windows under their own roof.

Brick buildings of the nineteenth century had solid, not cavity walls, and as the bricks were soft and porous, cement render and sometimes paint afforded some protection against the dampness. Kitchen, servants' quarters and storerooms were still separate in early examples, but later internal kitchens and water-closets were introduced.

From about 1840 onwards, houses appeared in the Gothic Revival style, asymmetrical in plan, with walls of stone or unrendered brick, steep slate roofs, clustered chimneys and carved and shaped bargeboards with long finials.

Weatherboard houses with corrugated-iron roofs were built in increasing numbers after the population rise of the gold rush days. In plan, one room usually projected at the front while a veranda ran along the front of the other room and sometimes along the side. Roofs were generally hipped, with a gable over the projecting room. A new feature was the use of curved corrugated iron for the veranda roof, the sheets either bullnose pattern with the curve at one end or slightly curved for their full length and laid with either convex or concave face upwards, or with ogee reverse curves at each end. Windows were hooded with iron or shingles. Light- or dark-stone paint colour was favoured for the walls, red oxide for roofs and wide stripes of red and stone colour for veranda roofs.

Cast-iron decoration was a feature on many nineteenth-century houses, as well as on some commercial and public buildings (Fig. 7.85). This material was screwed on where required to form friezes, enclosures, balustrades, fences, brackets, veranda posts, gates and garden furniture. At first imported from England, cast iron was produced locally in increasing quantities after 1860 with new patterns based on Australian motifs of kookaburras, cockatoos and lyrebirds, as well as fruit, fern and floral designs. It is interesting to note the current revival of cast decorative material, especially for garden furniture, though now aluminium and plastic are used as well as iron.

Figure 7.85 Imperial Hotel, Armidale, NSW; notable for its cast-iron decorations and typical of many country hotels of Victorian times, with its wide verandas, cement-rendered parapet, pediments and urns

Commercial buildings

Hotels in the early part of the eighteenth century were normally of two or three storeys, with wide verandas to each storey on the street frontage, often featuring cast-iron posts and decorative 'lacework' (Fig. 7.85). In the late Victorian period, many were built in the richly decorated Classic style and some, though considerably altered, are still in use.

Before the advent of large department stores, shops were single-storeyed of timber or brick. Heavy timber posts on the footpath supported a skillion awning, and a highly decorated parapet above hid the main roof. The shops were generally notable for poor internal lighting, scarcely mitigated by the skylights

that were often installed in the absence of side windows and not aided at all by dark-painted interiors and inadequate display windows.

Taller buildings were made possible by the introduction of lifts, which began to supplement stairways in Sydney in the 1880s, and office blocks, banks, and wholesale, retail, and bulk stores were now erected several storeys in height. Typically in the Classic style with the rich ornament of the period, many of these buildings are still used.

Public buildings

Churches were at first simple buildings, like the oldest existing church in Australia at Ebenezer, New South Wales. This was built in 1807 to the plan of an unknown designer and is still in use.

Well known among the Georgian churches are St Matthews in Windsor, St James in Sydney and St Lukes in Liverpool, all designed by Greenway. Another example is St Johns, Parramatta, whose twin towers were designed by John Watts in 1819.

Between the late 1840s and the 1890s, many churches were built in and around Sydney and

Figure 7.86 St Peters Church, Armidale, NSW; typical of Gothic-style churches of Victorian era; decoration, less than on cathedrals, is mainly on ends and entrance; note square belfry and difference in windows on entrance, belfry and end of building

in the larger country towns in the Gothic Revival style, usually large buildings with tall steeples and spires (Figs 7.72, 7.73, 7.76).

Most churches were of stone, but occasionally of brick, as in the three Greenway examples, and the early shingle roofs were later replaced with slates or copper. Pews and other interior furniture, fittings and joinery were of cedar.

In some cases a rectory was also provided (Fig. 7.87) and some are still occupied. A notable example is St Matthews rectory at Windsor, whose architect is unknown.

Figure 7.87 Rectory, Raymond Terrace, NSW; built 1840; similar in style to many country houses, with veranda on front and sides, and rather heavy veranda posts; stone walls and brick chimneys; break in line of veranda and main roof is rather unusual in this type of building

Post offices built during the period 1865 to 1890 range from the modest suburban type to the ornate General Post Office in Martin Place, Sydney. Most are still in use after a number of alterations and extensions. They were built in the Classic style in two storeys, often with a clock tower rising well above the main building. Walls were stone or cement-rendered brick, or brick with stone or cement-rendered string courses.

Town halls generally combined a large public hall with municipal offices. Most were built between 1880 and 1890 in the Classic Revival style, in stone or, more commonly, cement-rendered brick (Fig. 7.88).

Figure 7.88 Town Hall, Mudgee, NSW; built 1888; Greek Revival; brick walls, with cement rendering on front for contrast, triangular pediments over all windows, semicircular fanlight over entrance door; pediment, small tower, urns and circular louvres typical of Victorian architecture

Courthouses, too, are largely of the period 1880 to 1890 and were Classic in style built in brick or stone with slate roofs (Figs 7.77, 7.89). There are earlier examples at Hartley and Berrima, New South Wales (Fig. 7.75).

Schools were few in number prior to 1880, when the Education Act made attendance compulsory, but many were built during the next ten years, some of them still used.

Figure 7.89 Courthouse, Armidale, NSW; Greek Revival style; Ionic columns half-fluted, and square columns at either end; decorated clock tower and pediment typical features of this type of building, as are cement rendering and low windows on either side of entrance

Decorated walls of stone or brick with stone quoins and string courses were surmounted by slate roofs. Desks, fittings and joinery were usually of cedar.

Bridges were essential structures in the early days of settlement, for without them roads could not be extended or developed. Two in New South Wales are worthy of particular mention since their designer and builder, David Lennox, was Australia's first true bridge-builder. Both are of stone. Lennox Bridge, built in 1833 on the old road between Emu Plains and Glenbrook, is the oldest on the Australian mainland and was still usable until 1965, when it was declared unsafe for traffic. Lansdowne Bridge on the Hume Highway near Liverpool was built in 1836. With its fine elliptical arch, it still carries Sydney-bound traffic, while a new concrete bridge serves outward-bound travellers.

CONTEMPORARY BUILDINGS

Two important events retarded twentieth-century building in Australia: firstly, the Depression of the 1930s, when no money was available for building; and secondly, World War II in the early 1940s, when all labour and material were channelled into the war effort. These two events caused a shortage of all types of buildings and thus contributed to the postwar building boom. Another factor was the alteration of building regulations to allow greater scope for architectural design, particularly the increased allowable heights for buildings.

In the study of buildings, it must be remembered that interchange of ideas is worldwide; hence contemporary design is international, adaptions being made as necessary to suit local requirements.

Contemporary buildings of all types—domestic, commercial, and public—are quite different from their older counterparts. In New South Wales, changes in planning and

design have come about not only through changes in general concept of style, but also through a reorientation of attitudes to the siting of the various types of building. Thus new buildings in inner city areas are mainly office blocks, while homes, retail stores, and factories are moving out into new areas where land is more readily available and problems of transport and parking are minimised.

By examining each building type, its problems and requirements, it is possible to obtain a good overall picture of the changes that have taken place since the early years of the twentieth century and some of the reasons for them.

Domestic buildings

Houses show great diversity and imagination in design, resulting largely from efforts to make full use of the contours and natural resources of the site combined with skilful exploitation of the possibilities inherent in lower ceilings, flat roofs and split-level layouts.

In the older subdivisions, narrow, deep, rectangular allotments demanded a narrow house if a garage and driveway were required. In contrast, the newer areas offer wider, less regularly shaped blocks, permitting a layout in which the garage is included as part of the main building, thus providing a broad front elevation and easy access from garage to house (Fig 7.91). Fashions in the design of certain parts of the house change every few years. For example, the current choice is for hipped roofs; gables, if they are used at all, appear on the side elevations seldom at the front (Fig. 7.91). Brick veneer, an outer wall of bricks covering a timber frame, is a widely used construction technique, and colour contrast between walls and roof is often sought (Fig. 7.94). Decorative features such as veranda posts and brackets seldom appear.

Flats and home units offer high-density housing. A number of separate units, each with easy access to its own entrance, is planned in one block, with carpark, lawns and gardens an integral part of the layout. 'Town houses' or 'villa residences' are a recent development. These are grouped on the one site, but not in high-rise blocks. Local council regulations restrict flats and home units to certain specified locations and also govern the size and height of the buildings in relation to ground area, the provision of parking, garden and clothes-drying space, lifts and access to streets.

Figure 7.92 Modern double brick cement rendered home.

Figure 7.90 Trees left standing create play of light and shade, adding to overall effect of design

Figure 7.93 Fibre cement building, rendered with a special plastic coating to cover joints. Note sail type shade in the foreground.

Figure 7.91 Tudor-style architecture combines fibre cement sheeting with stained timber battens

Figure 7.94 Mansard type roof extension with cement fibre facing. Contrasting coverstrips give Tudor effect.

Figure 7.95 Copper chrome arsenate impregnated post and pole foundation to western red cedar weatherboard cottage. Pole construction is suited to sloping site

Figure 7.96 Prefabricated modular corrugated steel house design can be delivered anywhere in Australia. Victorian builder Reiny Loeliger

Influence of materials on style. In domestic architecture, as in other forms of building, new materials and techniques are often reflected in style changes. A summary follows:

Foundations—*Reinforced concrete footings, and slabs covering the whole site.*

Walls—*Welded structural steel; preformed prestressed and lightweight concrete; concrete bricks in modular sizes.*

Roof framing and structures—*Frames and trusses built up of steel or wood; laminated timber members; gang-nailing.*

Roof coverings—*Aluminium, galvanised iron and fibre cement (Fig. 7.96), all in new sectional shapes and lengths; guttering of similar materials in long lengths, with new joining and fixing methods.*

Figure 7.97 Combination of rough sawn timber wall, exposed beams and prefinished plywood ceiling accents open planning design

Figure 7.98 Ceiling follows rake of roof, with plywood lining panels in exposed framing; note solid timber wall panelling and open planning of living area

Wall coverings—*Concrete facings, preformed or in sheets for external and internal walls; fibre cement, hardboard, plywood, aluminium and weatherboards, in new sectional shapes and sizes for fixing horizontally or vertically (Fig. 7.95).*

Floors—*concrete slab as base; light-section wood coverings; composition coverings glued to concrete or wood.*

Doors and windows—*Wide, low designs in aluminium; flush and concertina-type internal doors; new sliding-door equipment; improved methods of hanging and securing doors and windows.*

Figure 7.99 Entrance becomes design feature of house and glass surround highlights prefinished plywood flooring.

Bathrooms and kitchens—*Wall facings in plastic laminates, fibre cement, hardboard, metal and glass; fibreglass, ceramic and glass built-in shower recesses; improved fittings.*

Lighting—*Indirect, recessed, fluorescent and multi-point systems (Fig. 7.99).*

Heating—*Central heating and wall-type heaters fuelled with oil, gas, or electricity.*

Ventilation—*Complete air-conditioning systems; fan systems of many types.*

Glass—*Not only admits natural light, but becomes a decorative feature in dividing walls and in bathrooms and kitchens; new sealers and adhesives have widened the scope of use.*

Adhesives—*Development of many new types with special properties has made possible innovations in finishing treatments for many kinds of surfaces.*

Paints—*Types for all surfaces, mostly plastic-based; simple and economical application; fast colours; washable films; resistance to most agents which hitherto affected paint adversely.*

Commercial buildings

The chief requirement for a modern office block is that it shall provide a number of separate rooms or suites of rooms, where business can be transacted separately and privately. Each office needs its own entrance, natural or artificial lighting and ventilation, and ready access from corridors served by fast, direct lifts. Display space for goods is seldom required, though advertising display space may be needed in individual areas. Planning must aim to make maximum use of all available internal rentable space so that financial returns on the buildings will be adequate. Provision must be made for vehicular access and loading docks for the delivery and removal of furniture and equipment. Parking facilities should cater for tenants' personnel and staff cars, but need not be so extensive as, for example, those of a shopping centre. The number of floors is unimportant since people do not need to move freely throughout the building in large numbers all at the same time, as in a department store.

These considerations explain why the typical modern office block (Fig 7.100) is a multi-storeyed building occupying maximum site area. Such tall, angular buildings are usually designed on the modulus of one floor, or even of one window-opening. Strong horizontal or vertical structural lines give them character, but the skilful designer will break the monotony of repetition with a curve or a change of angle, producing a play of light and shade. The main elevation may be given distinction by an imposing entrance, a piece of sculpture, or variety in materials or colours.

Retail houses have very different requirements. They need large, open spaces for displaying goods, room for many people to move about freely, ample customer parking and easy access from building to parking space and to public transport.

Shopping centres, now built away from the crowded inner city, are usually designed around a core of one or more departmental stores. They include a variety of smaller shops, and sometimes a hall or theatre. They are normally limited to three storeys, since transporting customers from floor to floor is costly in machinery and space. There is usually a central court or courts with access ramps, stairs, escalators or lifts serving all levels. Fountains, statuary and gardens may be integrated in the layout as decorative features.

Hotels and clubs are also moving to outer metropolitan areas. Hotels, with less emphasis on residential accommodation, are not multi-storeyed, but spread over the site to provide amenities and large entertainment areas, parking facilities, and easy access from streets and public transport. Good natural and artificial lighting and air-conditioning are important considerations. A landscaped garden or some piece of decoration provides a focal point in the layout.

Figure 7.100 The Riverside Centre, Brisbane—note unusual shape of building adds variety and interest to overall design.

Figure 7.101 The Regent Hotel, Sydney— strong vertical lines supported by a pyramid-like base.

Factories also are now built in suburban and rural areas, where there is more space for expansion and where certain land is zoned for industrial development. Factory buildings tend to expand outwards rather than upwards because transport from floor to floor takes time and space. Easy access to transport and good roads for the receipt and despatch of goods have a bearing on siting, while planning must allow for adequate staff recreation space and amenities. Modern factories are generally attractively designed with landscaping and gardens a feature.

Motor service stations have standard requirements such as easy access from streets, covered petrol service areas, workshops with built-in equipment, spare-parts storage and selling facilities, customer toilet amenities and adequate parking space. Service and parking areas are concrete-paved. Advertising and name signs are included in the overall design.

Public buildings

Contemporary churches differ vastly from older ones in the accepted Gothic style, retaining only the traditional steeple or spire and bell-tower, and even these show entirely new treatment. The chief requirement is that the main congregation space be well-lit, well-ventilated and acoustically satisfactory. Since the roof is usually also the ceiling, roof beams and lining become a special feature. Windows, doors and interior decoration often show traditional or historical influence. Entrance porches with room for people to congregate, landscaping, gardens and even parking facilities are practically standard in modern church design.

Government and civic buildings may number among their requirements offices, courtrooms and similar administrative space in constant use, and meeting rooms and entertainment halls which will be used only intermittently. A feature is made of interior lighting and decoration. For halls and entertainment areas, entrance foyers are scaled to allow for congregations of people and external approaches are usually fairly elaborate and planned for proximity to parking facilities and public transport. Extensive parking facilities are as a rule neither justified, owing to the intermittent demand, nor practicable, for reasons of space.

Theatres for 'live' performances have rarely been built in the twentieth century, but the period between the world wars saw the erection of many picture theatres. Some of their designs were produced in America and they run the gamut of architectural styles influenced by the flamboyance of the movies themselves. Decoration inside and out is lavish—lighting is extravagant and designed to produce an atmosphere; walls and ceilings are plastered, painted, gilded and ornamented—there must be no undecorated surfaces. It is all theatrical effect, in which every part of the building as well as stage, footlights and curtains has to play its role.

Schools have departed from the old concept of a single, multi-storeyed structure and are now built in several blocks or wings linked by covered ways, corridors and stairways giving access to rooms internally. Lighting and heating are essential factors in the overall design and sportsfields, play areas, gardens and parking are treated as integral parts of the layout.

New hospitals were few until after the Depression in the 1930s, when the modern concept of functional hospital design began to be publicly accepted. Prior to this many old buildings were altered and adapted to meet hospital requirements, but the new buildings were multi-storeyed with ample lifts, wide verandas, plenty of windows and special floor treatments to facilitate hygienic maintenance and minimise noise. Layouts provided for separate medical wings, accident and outpatient departments, with easy access for emergency services.

QUESTIONS

1. Describe what the earliest types of planes discovered were made of.
2. Discuss, in general terms, developments that have occurred in wood-turning lathes during the last 100 years.
3. Why was the first circular saw invented not a success?
4. Describe the wattle and daub construction used in the earliest building in Australia.
5. What influence did Francis Greenway have on early Australian Building?
6. Describe the influences concrete had on the style of buildings in the second half of the twentieth century.
7. Describe the space requirements of modern office buildings.
8. List the features that characterise modern-style furniture. Compare these with English period furniture of the latter half of the nineteenth century.
9. Name three English furniture designers, indicate the times in which they worked and discuss their particular styles.
10. Describe the work and some of the buildings of Edmund Blacket. Discuss his importance in relation to the second half of the nineteenth century in Australian architecture.
11. Name and describe a contemporary commercial building you have studied. Refer to its function; special features of shape, structure and decoration; external materials and finish; and artificial and natural lighting.
12. Selecting aspects of either buildings or furniture, discuss in general terms how trends in design are affected by: (a) the limitations and availability of traditional materials; (b) the development of new materials; (c) changing construction techniques; (d) sociological and economic influences.

SURFING THE NET

1. **Historical Buildings of Sydney**

   http://www.cityofsydney.nsw.gov.au/hs_historical_buildings.asp

2. **Blue Mountains Historical Buildings**

   http://www.mountains.net.au/history/structures.html

3. **History of Buildings at University of Sydney**

   http://www.usyd.edu.au/archives/smith.html

Chapter 8

The Australian Forest Industry

All Australians regard forest products—timber, paper, furniture, wood panels and so on—as an essential part of their lives. That is why the forest industry represents one of Australia's largest resources.

It has become Australia's second-largest manufacturing industry after the processed food and tobacco groups, with an annual sales turnover conservatively estimated at $10 billion.

There are two types of forests: native, mainly eucalypt hardwoods; and exotic soft woods, mainly pine.

The Australian Forestry Council has categorised Australia's forest resources primarily on potential commercial wood production criteria. It has identified 34.3 million hectares of native forests and 905 000 hectares of plantation forests. It did not count 6.5 million hectares of tropical eucalypt and paperbark forests in northern Australia, or 160 million hectares of woodlands.

About 75% of Australia's forest area is publicly owned (Crown land), including 25.6 million hectares of native forest and 0.63 million hectares of plantation. The remaining forest area is mainly privately owned native forest (8.8 million hectares), while a small proportion (280 000 hectares) consists of private plantations (Fig. 8.1).

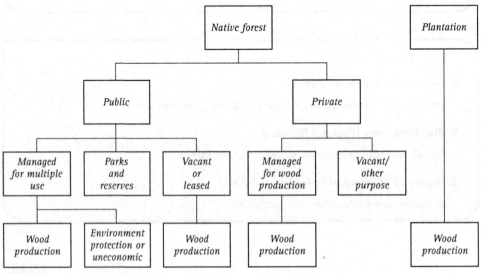

Figure 8.1 Diagram showing hierarchy of basic forest classifications

Less than 30% of the 25.6 million hectares of publicly owned native forests is managed for wood production. Only 1% of this managed portion is harvested in one year and this is immediately replanted with more trees.

Australia imports about one-third of its forest products.

About two-thirds of the logs removed from forests in Australia are from native hardwood forests; the remainder come from softwood plantations. Hardwood and softwood are not perfect substitutes in all end uses, each having different technical and economical advantages and disadvantages. Therefore, substituting softwood for all Australian hardwood production would not be feasible. A major reduction in the availability of Australian hardwoods would be reflected in higher hardwood imports.

COMPLEMENTARY USES OF FOREST RESOURCES

To meet the variety of consumer needs, the industry is made up of a number of sectors, including hardwood and softwood sawn timbers, pulp and paper, reconstituted wood and panel products and furniture.

The veneering, sawmilling and pulpwood sectors compete for wood resources at their margins; however, the bulk of the resources required by each are distinctive and, in most cases, regulated by allocation as well as price (Fig. 8.2). Moreover, pulpwood industries provide a market for waste material from sawn timber production. Wood is used for pulpwood when it is not suitable for sawing because of size, shape or defects (Fig. 8.3).

At present, most of the pulpwood harvested in Australia is exported, rather than being processed domestically. These overseas

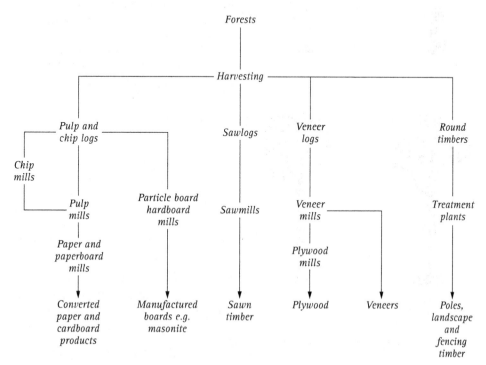

Figure 8.2 Forest product industries

industries use headlogs, pulp-logs, thinnings and sawmill residues to manufacture a wide range of products, from reconstituted wood panels to fine writing paper. (Thinnings are young trees cut to reduce the competition in a stand, allowing the remaining trees to grow to their full size. If headlogs, thinnings and other pulp are not removed from the forest floor, they add fuel to any fires that may occur.)

Forest management is a practice that has come under close public scrutiny in the past decade. Such scrutiny and the growth in lobbying power of the conservation movement have contributed towards the creation of a set of rules known as 'Codes of Practice in Public Forests'. Forest management involves integrated harvesting in native forests (where both sawlog and pulpwood are harvested in the one operation) within the framework of the Code, which calls for the exclusion of harvesting from such areas as streamside reserves, catchment areas and native animal habitats.

The Forestry Commission says the volume of timber is much greater from an integrated management system, in both the short and long terms. According to the Commission, the system allows for regeneration similar to that which occurs during the natural forest cycle.

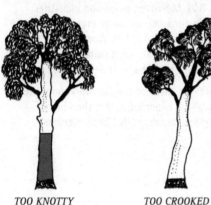

TOO KNOTTY
(HEADLOG)

TOO CROOKED
(TOO DIFFICULT
TO SAW)

TOO DEFECTIVE
(FIRE SCAR,
DECAY ETC.)

TOO SMALL
(HEADLOG)

Figure 8.3 Residual wood left after sawlogs have been cut from trees of various shapes and sizes

HARDWOOD PLANTATIONS

Hardwood is mainly derived from native forests and accounts for two-thirds of the wood removals in Australia. Of these, about 60 000 hectares of hardwood plantations are managed mainly for pulpwood. At present, harvesting from these areas is limited.

Eucalypts take about 80 years to grow to a size suitable for harvesting. As a result, in the opinion of the National Association of Forest Industries, it is uneconomical to produce sufficient sawlogs from a hardwood plantation. The expansion of plantations needed to replace native forests as a resource would compete with agricultural and other land-uses, require an industry restructure and cause serious social dislocation.

The Association maintains that the replacement of native forests by plantation is unnecessary, as the non-wood values—such as recreation, water catchment, soil preservation and the conservation of forest-dependent animal and plant species habitats—can be protected by careful logging techniques.

REGULATIONS AND HARVESTING PLANNING

The following list of regulations apply to logging operations:

- Code of logging practice—State-wide conditions that outline procedures for licensing, operations, legalities, broad environment conditions and occupational health and safety.
- State licence conditions—set out procedures to be followed by individual operators and customers.
- Codes of procedure—ensure revenue is collected for all material removed from the forest and that logs are used for the highest value of products.
- Harvesting plans—site-specific documents that describe each logging area and the necessary environmental safeguards. The nature of the forest, aims of the operation and constraints are detailed. Harvesting plans set out when, where and how the operation is to take place.

Harvesting plans contain conditions that are designed to supply timber whilst minimising environmental effects of logging by:

- Restricting the area to be disturbed to suitable locations and defining which trees will or won't be felled.
- Ensuring protection of the soil through erosion control.
- Retaining wildlife habitats, and maintaining undisturbed vegetation around streams, drainage lines and wildlife corridors.
- Rehabilitation of logging sites.
- Avoiding environmentally sensitive areas. These may include: habitats for endangered species, rainforest, steep terrain, archaeological sites, scenic areas or recreation areas.

THE "LOGGERS"

Logging contractors are paid by the sawmill, usually on a log volume basis. Each logging contractor has a Contractor's Licence which authorises their company to work and remove timber from State Forests. The workers in turn, must have Operator's Licences. To obtain an Operator's Licence a worker must be proficient in the safe use of the relevant equipment, and accredited by a State Forests instructor.

The logging crews are supervised by State Forest field staff to ensure they are complying with operating conditions. These are designed to protect the environment.

USES OF HARDWOOD IN AUSTRALIA

A much lower proportion of hardwood removals are used for sawlogs and veneer logs, compared with softwood. Hardwood is mainly used for sawn timber and pulp and paper manufacture (Fig. 8.4). Hardwood is generally denser and stronger than softwood. A number of hardwood species exhibit natural durability without the need for preservatives. This makes them ideal for use in situations where resistance to biological degradation is important, for example, civil engineering.

SOFTWOOD PLANTATIONS

Softwood harvesting in Australia accounts for about one-third of the total wood removals and is growing in importance. Most softwood produced in Australia is from plantations, although there is some harvesting taking place in native forests.

Plantation softwoods in Australia have a comparative advantage over native forests because of their geographic concentration, higher productivity per unit area and the consequent reductions in harvesting and transport costs. There is also a much greater degree of resource security in the softwood sector.

Softwood sawlogs take 25 to 40 years to mature (compared with 80 years for hardwood). It is important to note that all the wood grown is fully utilised. A minimum amount of raw material must be available for

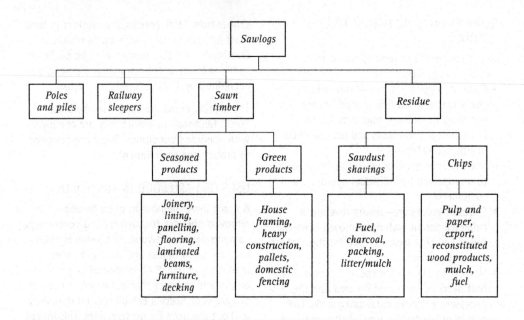

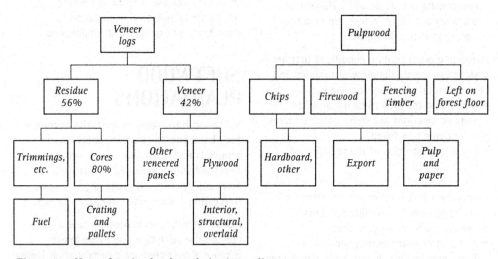

Figure 8.4 Uses of native hardwoods in Australia

USES OF SOFTWOOD IN AUSTRALIA

Softwoods have to be kiln-dried for strength and stability, providing an additional cost to the sawmiller. Also, the knotty appearance of the establishment of facilities for processing thinnings and sawmill residue.

the softwood produced in Australia is not valued highly by domestic users. This has led to the concentrated use of softwood in the structural market for sawn timber; for instance, in house framing. It is also used extensively for core material in plywood, reconstituted wood products and in the manufacture of pulp and paper (Fig. 8.5).

PRODUCTS STEMMING FROM NATIVE FORESTS

Native forests support a number of primary industries within them, providing a suitable site for grazing and the production of such diverse products as honey, oil and firewood, while timber taken from the forests is used in a number of industries to create a wide range of products.

Native timbers have become popular with many facets of industry. Sawn timbers are converted into many different forms for domestic use. They are used as stress-graded beams for building construction, laminated beams and laminated timber bench tops as well as in more specialised fields, such as woodturning and boat-building (Fig. 8.4).

Sawmilling products—*Sawmilling involves two separate approaches—mills close to the source of the raw material and those located in large towns and cities that re-saw timber. Native forest products are consumed in the rough-sawn state as bearers, joists, fencing and stakes (for agricultural industry pursuits such as horticulture and the cultivation of oysters), or are further processed by various manufacturing interests within the furniture and building trades. Many of the city and town mills further process their timber into various boards for the building trade, including flooring, skirting and architrave mouldings, sills, door jambs and so on.*

Veneer—*Provides the base product for both the plywood and panel industries, where veneers are built up into panels, boards or facings for solid-core stock, medium density fibreboard or particle board. Veneer manufacture usually includes all the processes for further processing into plywood.*

Poles, sleepers and mining timber—*These products come from timber which often needs to be treated with preservative and, depending on the species, needs to be de-sapped to reduce borer susceptibility. The sleeper-cutting industry utilises a number of eucalypts including ironbark, tallowwood, greygum, greybox, white mahogany and river redgum. Waste from these practices is used to supply charcoal for industries using fuel-fired boilers and the domestic market (via barbecue fuel). The mining industry utilises both round and sawn timbers.*

Craft products—*As traditional species such as red cedar and silver ash become scarce, the craft industry is turning to alternative sources of outstanding colour and figure. Because of their natural beauty, Australia's native species are becoming very popular with craftspeople.*

Landscaping materials—*A general awareness that the exterior of the home demands as much attention as the interior has led to a boom in landscaping materials. Unusable material from the sleeper-cutting industry, for example, is sawn into planks, undersized sleepers and posts; the tops, branches and other discarded forest material is used as chipped mulch for the horticulture industry; and some of the waste from mills is collected to manufacture hardwood sawdust compost as well as hardwood chips.*

Posts and poles (for domestic use)—*These products are used in the building of pole-frame constructions, as veranda posts and fence posts in stockyards, for other rural uses and as raw material for the production of compost. Poles or posts destined for structural use are treated to prevent pest attack and extend the life of the timber. The treatment involves the use of chemical preservatives, which are applied under a closely controlled vacuum/pressure process.*

Hardboard—*Produced from hardwood chips, which are broken down and pressed under steam and pressure to form a board.*

Pallets—*The pallet industry is growing as more and more of our domestic products are handled in bulk by machinery.*

Pulp—*Can be made from chips of timber sourced from softwood plantations and native forests. It is used as a basis for paper making.*

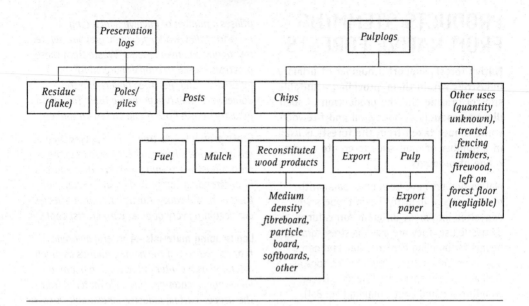

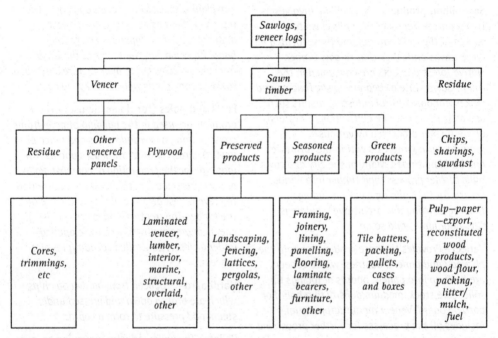

Figure 8.5 Uses of softwoods in Australia

Paper products form a very important part of the end use of timber resources. Paper products also include corrugated and solid board containers and cartons.

Furniture—In Australia, the furniture industry is very diversified, ranging from one-person operations making single, quality items of cabinet construction, garden furniture, turnery and suites to very large and complex factories using every conceivable style and construction method to produce furniture for domestic and export markets. The furniture manufacturing industry uses natural timber, panels (such as MDF and particle board, plywood and hardboard) and veneers for facing as well as imported timber.

Building materials—If both domestic and commercial structures are taken into account, the building industry is the greatest user of timber in Australia today. Timber in every shape and form is used on every building site. Even when a structure uses reinforced concrete and steel frames, it would almost certainly use timber formwork and timber would be used for doors and trimmings.

PRODUCTS STEMMING FROM PINE PLANTATIONS

Timber taken from the pine plantations serves a number of industries and is used to produce a range of items (Fig. 8.5).

Landscaping materials—Derived from two sources within a pine plantation: the bark from large logs and the trimmings, which are converted into pine chip, or pine flake for use as mulch.

Softwood chips—Produced from pine logs in a number of stages of the life cycle of the plantation. They are used in the production of particle board or medium density fibreboard.

Manufactured boards and panels—The range of boards being produced from softwood is increasing. It includes medium density fibreboard and particle board.

Veneer—Produced from radiata and hoop pine, and used in the production of plywood and face veneer particle board.

Sawn timber—Can be used in the rough-sawn state or treated to prevent decay and insect attack and then used in a number of domestic and commercial situations. Sawn timber may also be dressed or milled for use in the cabinetmaking and carpentry fields. The timber can be stress-graded for structural use, such as wall framing and roof trusses.

Residue—Sawmill residue is used as a raw material for the production of compost for the horticultural industry and waste is chipped for the pulp and paper industries.

Fencing—The popularity of pine as a fencing material is increasing because of its light weight for transport and cheap machining (for such uses as picket-profiled fencing). When treated it has a long life span.

Poles—Produced at a number of stages in the life cycle of the plantation. They are generally treated and used in the building of pole-frame constructions, as veranda and fence posts and in the building of retaining walls.

Imports—These products play a very important role in the Australian timber industry, accounting as they do for more than 20% of the timber needs of industry.

Exports—Fall into a number of categories including woodchips, rough-sawn logs, fence posts, veneer, sawn-wood, chemical wood pulp, paper and paperboard. Hardwood chips account for 85% of the forest industry's total exports. The total value of exported products in 1986–1987 was 1.1% of all Australian exports.

THE ENVIRONMENT AND THE FOREST INDUSTRY

The relationship between the awareness of the need to conserve native forests for environmental reasons and industrial activity has been a stormy one. The conservation movement argues for the halt

of logging and the extension of national parks, while the forest industry argues that forests are resources to be harvested. The confrontation between the conservationists and the pro-loggers has led to more recognition of the importance of forests as wildlife habitats and storehouses of information about animal and plant species.

The Forestry Commission points to its Code of Management, which outlines the dos and don'ts of logging, as providing a check against logging practices that are damaging to the forest ecology. The Commission's foremen are schooled in the art of identifying various habitat trees and areas, with the aim of ensuring special care is taken not to disturb those habitats.

The Commission practises a system of felling known as selective logging. Forests in the north-eastern corner of New South Wales have been selectively logged for more than 100 years. Cutting is carried out on a cycle over several years, with only a limited number of trees felled per hectare. The end result is a forest of large trees, some medium height and others in their infancy. The Commission maintains that selective logging is actually mimicking nature.

In the logging versus conservation debate, emotions run high. The controversy has risen above its own arena to become a political question. At stake, say those in the industry, is the direct employment of 100 000 and indirectly for up to 200 000, while the opponents of logging say that at stake are the diversification and continuation of many plant and wildlife species.

THE GREENHOUSE EFFECT

Scientists have recorded that the composition of the atmosphere is changing. The amount of carbon dioxide, methane, nitrous oxide and chlorofluorocarbons (CFCs) is rising. The causes include our increasing population and the matching increases in motor vehicle exhaust emissions and use of fossil fuels. Deforestation and burning off are also significant sources of carbon dioxide emissions.

A summary report, issued in August 1990 by the Australian and New Zealand Environment Council, describes two of the major 'sinks' that remove carbon dioxide from the atmosphere as the world's forests and oceans. The report states that protecting and retaining existing tree cover and increasing the total forested area of Australia must be a high priority.

The Council, made up of government representatives from the two countries, is aiming at coming up with a national strategy in light of the greenhouse predictions of changing climates and sea levels. It states that afforestation (planting trees on cleared land) in the form of plantations, and encouragement of re-growth, will help slow down the rate of carbon dioxide build-up in the atmosphere.

However, the Forestry Commission's Wood Technology and Forest Research Division says its latest findings indicate that mature forests produce as much carbon dioxide as they absorb. Therefore, it says, mature forests or 'old growth' forests are responsible for no net absorption of carbon dioxide. Studies have shown that this is because unlogged forests are in a state of dynamic equilibrium, where the decay rate matches the growth rate. As a result, the Commission says, the emission of carbon dioxide and other gases could be eased by the continuous logging of forests.

In considering future options, the government will have to take into account the survival of the forest industry and the needs of consumers of forest products while considering the impact of logging on the forest ecology and the pressing need to preserve Australia's natural heritage.

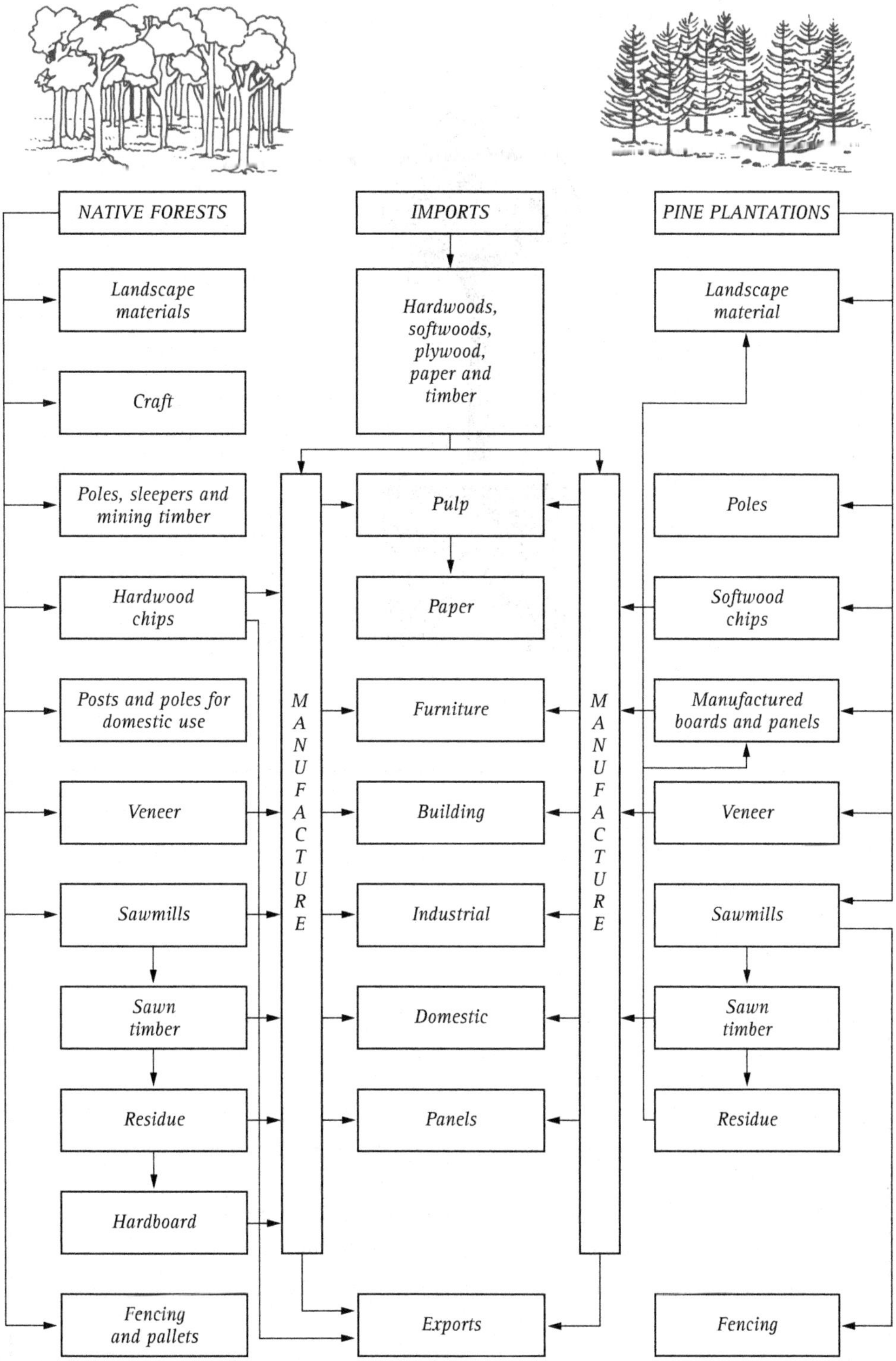

NATIVE FORESTS	IMPORTS	PINE PLANTATIONS
Landscape materials	Hardwoods, softwoods, plywood, paper and timber	Landscape material
Craft		
Poles, sleepers and mining timber	Pulp	Poles
Hardwood chips	Paper	Softwood chips
Posts and poles for domestic use	Furniture	Manufactured boards and panels
Veneer	Building	Veneer
Sawmills	Industrial	Sawmills
Sawn timber	Domestic	Sawn timber
Residue	Panels	Residue
Hardboard		
Fencing and pallets	Exports	Fencing

MANUFACTURE

Figure 8.6 Forest-based industries

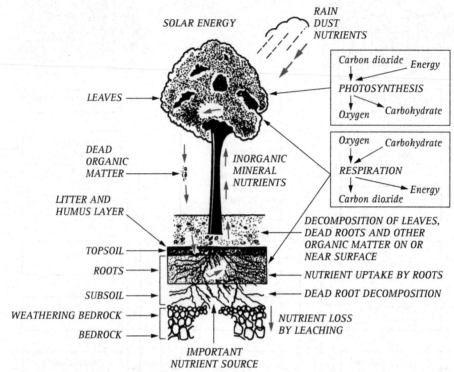

SOLAR ENERGY

RAIN
DUST
NUTRIENTS

Carbon dioxide — Energy

PHOTOSYNTHESIS

Oxygen — Carbohydrate

LEAVES

Oxygen — Carbohydrate

RESPIRATION

— Energy

Carbon dioxide

DEAD
ORGANIC
MATTER

INORGANIC
MINERAL
NUTRIENTS

LITTER AND
HUMUS LAYER

DECOMPOSITION OF LEAVES,
DEAD ROOTS AND OTHER
ORGANIC MATTER ON OR
NEAR SURFACE

TOPSOIL

ROOTS

NUTRIENT UPTAKE BY ROOTS

SUBSOIL

DEAD ROOT DECOMPOSITION

WEATHERING BEDROCK

BEDROCK

NUTRIENT LOSS
BY LEACHING

IMPORTANT
NUTRIENT SOURCE

Figure 8.7 The mineral nutrient cycle

QUESTIONS

1. What percentage of Australia's publicly owned native forests are managed for wood production?

2. What percentage of managed public native forests are harvested per year?

3. What percentage of its forest products does Australia import?

4. Compare the percentage of hardwood logs removed from Australian forests with the percentage removed from softwood plantations.

5. How would a major reduction in the access to or availability of Australian hardwood affect the Australian economy?

6. Explain the meaning of 'integrated harvesting' in forest management.

7. Why is it considered uneconomical at present to produce plantation hardwood sawlogs in Australia?

8. What are the non-wood values of native forests? How can they be preserved?

9. List the main uses for plantation softwoods grown in Australia.

10. List the main products derived from native forests in Australia.

11. What primary industries do our native forests support?

12. Why is woodchipping such a contentious issue in the community? What is at stake in this controversy?

13. What are the probable causes of the greenhouse effect? What effect could it have on our future, climatically and health-wise?

14. Explain why a young forest may help to protect us from the greenhouse effect.

SURFING THE NET

1. **National Association of Forest Industries**

   http://www.nafi.com.au

2. **State Forest of NSW**

   http://www.forest.nsw.gov.au

Chapter 9

Designing and Planning for People

The underlying purpose in preparing **Woodworking, Part One and Part Two** has been to do far more than simply provide a 'how to make' book with a collection of projects. Woodworkers should certainly possess a sound knowledge of wood and its products and a basic proficiency with tools—but it is also important that they have an understanding of the principles behind good design so that they are able to experience the pleasure of developing their own designs.

We have endeavoured to discuss in detail the principles behind good design and have attempted to illustrate these principles with examples of our own work and the work of recognised craftspeople, furniture-makers, teachers and students.

We have tried to provide sufficient information and know-how to encourage the reader to create his or her own designs, to suit their particular needs or tastes.

Applying the design process to satisfy a need is a more rewarding exercise than simply working to an existing plan. Modifying an existing design may be all that is required, but it can still be satisfying. This is often referred to as the partial or modified design process. The application of the design process to solve a problem or satisfy a need simply involves considering the various factors of this process to achieve a result that will be pleasant to look at and do its job well.

Designing a piece of furniture without any identifiable style is probably the most common mistake made by aspiring woodworkers. It is most important therefore to choose designs that have clearly recognisable styles. The idea is not necessarily to copy, unless an exact replica is required, but to incorporate styling features into the design so that the item fits in with the surrounding furniture.

ERGONOMIC DESIGN

The difference between a well-designed piece of furniture and one that is useless is determined largely by how the furniture fits the human frame and how it adapts to the way people use it.

Serious designers study the comparative dimensions of the human body, to arrive at the initial scale and dimensions of a piece of furniture. The science of anthropometrics standardises certain body measurements so that the designers can get an idea of the parameters imposed by the human frame on their designs (Fig. 9.1).

It is important to realise that the 'average physical proportions' used in furniture design satisfy the majority of the population but do not necessarily satisfy other people at the extreme ends of the size scale. Even though a designer tries to cater for at least 90 per cent of the population, the top and bottom five per cent are not always taken into consideration.

These limits are known as the fifth and ninety-fifth percentiles (Fig. 9.2). The designer should therefore further refine data to fit a particular case. Think of designing furniture for children for instance—how do you make allowance for their growth?

CHAIRS

When a person sits on a seat, the body's weight is taken off the legs and feet and is transferred to the buttocks and the backs of the thighs. When a person leans back against the backrest, it helps the spine maintain its natural curvature.

Seat

The seat height is most important—if the seat is too high, it causes uncomfortable pressure under the thighs and restricts the blood supply to the legs (Fig. 9.3) whereas if it is too low, the sitter assumes a crouched position (Fig. 9.4). The ideal height (approximately 480 to 500 mm) allows the sitter to have both feet flat on the floor, with a small gap under the thighs just behind the knees.

The seat should have a slight backward slope to help hold the person in the chair, but a desk chair seat is horizontal or only slightly tilted as the user needs to be more upright.

The width of a seat should accommodate the spread of the buttocks and additional room should be provided to allow for movement and thick clothing. The seat depth should allow for maximum body support—too shallow a depth of seat applies extra pressure to thighs and buttocks.

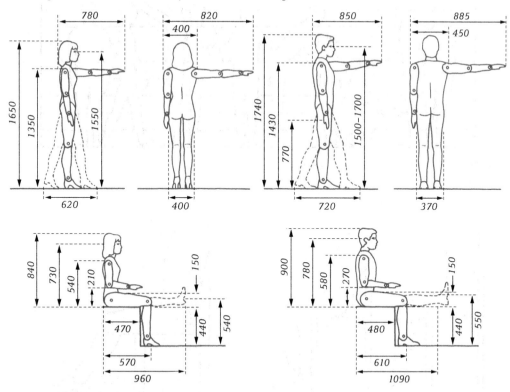

All measurements in millimetres

Figure 9.1 Diagrams indicating how average dimensions of men and women determine dimensions of furniture

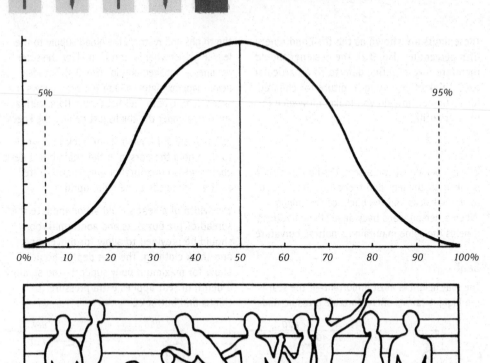

Figure 9.2 Graph showing distribution curve for given number of people in relation to height; some people are very tall and some are very small, but majority fall into area either side of middle or 'mean' depth

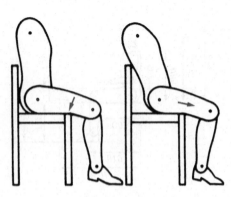

Figure 9.3 If seat is too high, it applies pressure underneath the thighs and causes person to slide forward

Figure 9.4 If seat is too low, person assumes cramped position and it becomes difficult to get out of chair

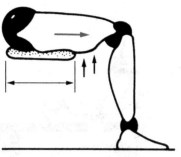

*Figure 9.5 Too shallow a depth of seat
applies extra pressure to thighs and buttocks*

The surface composition of a seat is also very important. It can be solid, smooth or shaped and covered with padding, cane or fabric. Obviously a person will be more comfortable in a soft seat but if the seat is too soft, the body's weight is distributed to the broad sensitive areas of the pelvis and thighs. For seating over a long period of time, the body should be supported under the centre of the pelvis and the padding should only depress approximately 12 mm. Carved and sculptured seats give the appearance of comfort but unless they are designed for a specific person's shape, the advantages are minimal. A textured surface can be an advantage as it stops the sitter from sliding and therefore helps maintain the correct seating position.

Backrest

The backrest should give support at the right points in the lumbar region (Fig. 9.6). It must be wide enough to allow room for the buttocks and the area just below the lumbar region and it should also allow room for the sitter to arch the back to relieve strain. Therefore the backrest should be between 300 and 350 mm wide. The tilt of the backrest helps to position the body comfortably and to prevent the user from gradually sliding forward but the angle must be suited to the way the chair is going to be used—either for work or relaxation. A tilted backrest becomes a disadvantage on chairs used for working, such as desk chairs, which are required to be more upright.

Armrest

Armrests are required to perform a number of functions. They provide body support and can be used for leverage when sitting down or getting out of the chair. The important dimensions for the armrests are the width between them, and the height or the distance from the seat (Fig. 9.7).

Types of chairs

The specific purpose of a chair dictates the degree of support to be given to the back, shoulders and neck.

The easychair should have armrests fitted, and the seat and the backrest should be tilted for comfort. The dining chair does not necessarily need arms, but it should be in a more upright position so that food is not spilled. The study chair needs to be well designed to suit the user, who generally spends an extended time using the chair and needs to be well supported so as not to become fatigued too quickly.

TABLES

Although there are accepted standard figures for the size of a table according to its use, the final dimensions for a design should be determined by the kinds of tasks to be performed on it and the number of persons who require access to it.

DESKS

The design of a desk must suit the user over long periods of time, under conditions that require concentration and energy and where unnecessary fatigue must be avoided. The height of a computer desk is determined by a number of factors, including the height of the screen and the height of the keyboard. A height of approximately 660 mm would help ensure that the user is seated correctly. There must be a number of storage areas, such as drawers and shelves, within reach so little effort is required to access materials.

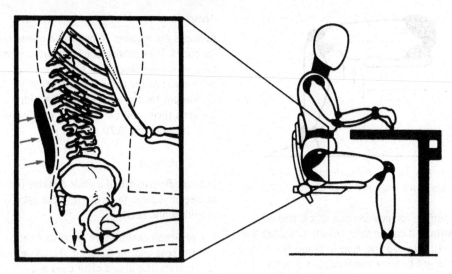

Figure 9.6 A correctly supporting backrest suited to use of chair

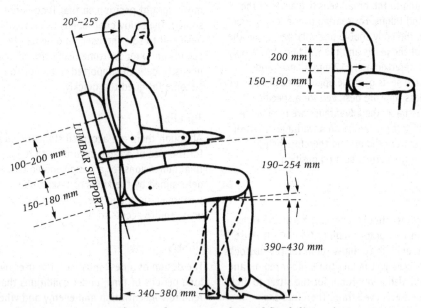

Figure 9.7 A person correctly seated at an armchair of fixed dimensions

DINING TABLES

A dining table is a very aesthetic piece of furniture and is often designed to suit a certain style of furniture already used in the house. The height of the table is determined by the need of the users to press down on their knives and forks to cut their food (approximately 710 mm). Eating is a social habit, therefore the width and length of a dining table should allow easy conversation plus enough room for people to be seated comfortably and not feel crowded (allow approximately 620 mm 'elbow room' for each person, increased to 900 mm at the corners of rectangular tables). An important design feature which is often overlooked is the placement of the table legs. These must be placed so as to allow people to sit without being restricted.

OCCASIONAL TABLES

Occasional tables are used to support coffee cups, magazines and so on or serve purely aesthetic purposes. They normally conform strongly to the existing style of furniture in a room. Depending on their use, they are approximately 300 to 500 mm high.

STORAGE UNITS

When designing a storage unit or cabinet, it is still necessary to consider the dimensions of the human body as these determine how easily a person can lift and manipulate objects. Lifting objects in and out of a storage unit should not cause excessive strain or have to be done at an awkward angle. Other factors that influence the design include the area needed around the storage unit to open it, whether or not the objects stored in the unit are to be visible (glass doors may be required), the frequency with which the objects are used and the size, weight and value of the objects.

DESIGN EXERCISES

Listed below are a number of design exercises that can be worked through to develop your skills; they include small, simple items and more difficult ones.

1. Design an item of furniture that would be suitable to support a computer and its accessories. Consider:
    - size of user;
    - type and size of computer and accessories;
    - style;
    - stability.

2. Design a suitable device to support a person while using a computer at the above piece of furniture. Consider:
    - size of person;
    - ergonomics;
    - strength (construction);
    - style;
    - comfort.

3. Design a frame to hold a mirror. Consider:
    - placement;
    - size of mirror;
    - style and proportions of frame;
    - method of fixing to wall.

4. Design an item of furniture to house and display china and glassware. Consider:
    - style (type of articles to be housed);
    - size (number and size of articles).

5. Design a set of containers suitable for use in the kitchen. Consider:
   - hygiene;
   - style;
   - function (ability to protect contents, type of thing to be stored).

6. Design an item of furniture to house and protect a stereo sound system. Consider:
   - size and type of equipment;
   - dust protection;
   - ventilation;
   - wiring;
   - style.

7. Design a pair of containers to hold and dispense salt and pepper. Consider:
   - size of containers;
   - size of dispensing holes;
   - style;
   - function (method of holding salt and pepper in and method of refilling).

8. Create a container to hold fruit. Consider:
   - size (amount of fruit);
   - stability;
   - type of timber;
   - finish.

9. Design an item of furniture for use in a bedroom to hold and organise small items of clothing. Consider:
   - style of bedroom furniture;
   - size (amount of clothing);
   - accessibility.

10. Design a suitable device to support a person while eating at a breakfast bar. Consider:
    - height of bar;
    - size of the person or people using bar;
    - ergonomics;
    - strength (construction);
    - style;
    - comfort.

GALLERY

The following projects are examples of woodcraft and furniture by students and leading Australian craftspeople. They are not intended to be a catalogue of ideas for the woodworker to copy; if this had been the intention, they would have been accompanied by working drawings with measurements. The projects are included purely as examples of fine quality work and design and as a starting point for you to create your own beautiful piece of furniture or woodcraft and to experience the supreme satisfaction to be derived from doing so.

Figure 9.8 'Petal bowl' turned and carved from jacaranda; inspiration derived from nature; designed and made by Les Fisher. Photo: Keith Jeeves

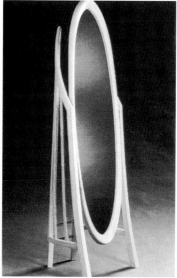

Figure 9.9 Cheval Mirror. Clean feminine lines enhance the beauty of silver ash (Flindersia bourjotiana). By Richard Vaughan; photo: Scott Donkin

Figure 9.10 Delightful jacaranda vase designed and turned by Keith Jeeves

Figure 9.11 Twin cylinder overhead camshaft engine in stacked plywood by Richard Crossland

Figure 9.12 Rustic jacaranda jewel box with satin finish by Leady

Figure 9.13 Unusual box made from silver ash and zebrano; note ebony hinge; by Ernie Pauchet

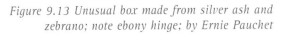

Figure 9.14 American walnut box with simple hinge arrangement by Ernie Pauchet

Figure 9.15 Sapele mahogany vanity mirror and drawer. Note book mark veneer on top of carcass with silver ash insert surround; by Al Usher

Figure 9.16 Red cedar hanging pomander by Lindsey Dunn. 700 mm × 150 mm

Figure 9.17 Rocking motor bike. A group project by members of the Oyster Bay Woodturners' group

Figure 9.18 Clock turned from Blakeley's redgum burl by Lindsey Dunn

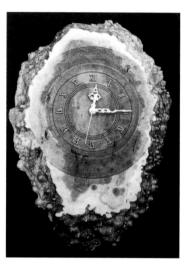

Figure 9.19 Rosewood box and drawers, inspired by nature; bandsawn, reassembled and hand carved from a single piece

Figure 9.20 Sound system cabinet, the floating top and space behind the drawers allow heat to escape. Designed and made from Queensland maple by Richard Vaughan. Photo by Greg Piper

Figure 9.21 Lidded box with interesting incised turning on lid and ends. Note simple hinge arrangement. By Ernie Pauchet

Figure 9.22 Small desk designed and made by Richard Vaughan from Sydney blue gum. The legs are laminated. Photo by Ian Hobbs

Figure 9.23 Boxes with lids designed and made by Roger Gifkins. Roger is recognised internationally for his beautifully designed boxes

Figure 9.24 Rosewood lattice bowl designed and turned by Lindsay Dunn. Circles are turned and radial cuts are made with a router

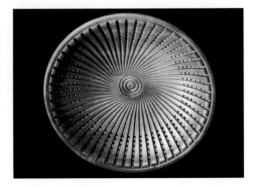

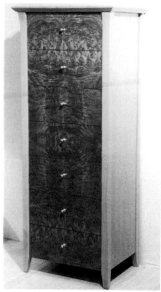

Figure 9.25 Set of drawers made from silver ash and European beech. Note bird's eye figure on face of drawers. By Covemore Designs

Figure 9.27 Segmented red cedar and silver ash turned box and lid; designed and made by Ernie Pauchet

Figure 9.26 Interesting chest of drawers made from the sapwood of camphor laurel. By Al Usher

Figure 9.28 Vanity mirror and drawers made from radiata pine

Figure 9.29 Lounge table designed and constructed from Sydney blue gum by Phil Newell. The table top is made from solid granite

Figure 9.30 Sydney blue gum hall table. The slender lines are achievable because of this timber's strength. The glass top is an attempt to capture the unique way gum-tree foliage dapples the light. The glass was slumpted by Warren Langley to a stylised leaf pattern carved on MDF. Designed and made by Richard Vaughan

Figure 9.31 Pisces table in Tasmanian blackwood. A wedding gift for a couple who met and fell in love by the sea. The legs are fish-shaped in section and the top has been textured with a carving tool to catch the light like moonlight on the water. Designed and made by Richard Vaughan

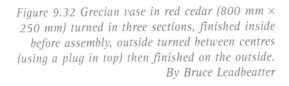

Figure 9.32 Grecian vase in red cedar (800 mm × 250 mm) turned in three sections, finished inside before assembly, outside turned between centres (using a plug in top) then finished on the outside. By Bruce Leadbeatter

Figure 9.33 Major project designed and constructed in solid silky oak by Joshua Rosenthal for the higher school certificate in NSW

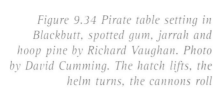

Figure 9.34 Pirate table setting in Blackbutt, spotted gum, jarrah and hoop pine by Richard Vaughan. Photo by David Cumming. The hatch lifts, the helm turns, the cannons roll

Figure 9.35 Tasmanian Oak cabinet in a country style designed and made by Michael Leadbeatter

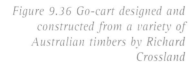

Figure 9.36 Go-cart designed and constructed from a variety of Australian timbers by Richard Crossland

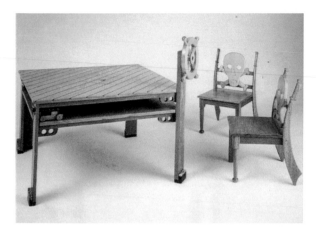

Figure 9.37 Hall table in red mahogany designed and constructed by Richard Vaughan. Photo by Greg Piper. The timber, carbon dated at about 8000 years old, was dug up during the excavations for Sydney's new southern rail link. The table was commissioned to commemorate the completion of the project. The legs refer to core samples and construction piles. The under profile of the top refers to geological strata

Figure 9.38 Lounge storage unit in English beech with bookmatch bird's eye maple veneered door panels by Covemore Designs

Figure 9.39 One of a set of carver dining chairs in Brazilian mahogany by Richard Crossland

Figure 9.40 Small table desk in Brazilian mahogany by Al Usher

SURFING THE NET

1. Ergonomics

http://www.ergonomics.com.au

2. Woodlink

http://www.vicnet.net.au/~woodlink

3. Australian Wood Artisans

http://www.woodart.com.au

4. Woodweb

http://www.woodweb.com

INDEX